INFORMATION TECHNOLOGY
A Finance Professional's Perspective

INFORMATION TECHNOLOGY
A Finance Professional's Perspective

Srinivasan M K

KNOWLEDGE WORLD
KW Publishers Pvt. Ltd.
New Delhi

in association with

way forward
Integrated Capital Services Ltd.
Member of
BTG Global Advisory

In association with

Integrated Capital Services Ltd.
www.raas.co.in

Member of
BTG Global Advisory
www.btgga.com

ISBN 978-93-83649-46-4

Published in India by Kalpana Shukla

KNOWLEDGE WORLD
KW Publishers Pvt Ltd
4676/21, First Floor, Ansari Road, Daryaganj, New Delhi 110002
Phone: +91 11 23263498/43528107
Email: knowledgeworld@vsnl.net • www.kwpub.com

2011 FPBA BEST PUBLISHERS AWARD (ENGLISH)

Printed and Bound in India

Contents

Section 5: Network, Internet and Telecommunication

Section 6: Business System

Section 7: Deploying Business Applications Software

Section 8: Access and Identity Management

Section 9: Advanced and Emerging Technologies

Dedicated to my parents, who are my first teachers!

Terms of Use

Acknowledgements

Writing a book for businesspeople and professionals has always been a dream for me as I thought that it is one of the easier ways for disseminating my knowledge and passing on my legacy to the next generation. The writing of this book is a first step in that direction.

In the journey of writing this book, there were several people who supported me in different ways. I think it important to thankfully acknowledge their support. My thankful acknowledgement is neither a ritual nor sentiment, but deeply soaked in my gratitude.

Firstly, my wife Vijaya who constantly encouraged me for writing this book by extending incomparable support in not only writing but also completing it successfully.

Then, my elder son Sumanth—a creative genius who does several activities from writing computer codes to reading to playing the guitar to composing music and to writing blogs. When I requested his help for reviewing this book for improvement, he accepted and completed it within the scheduled time in spite of his various professional priorities.

My younger twin children, Hemanth and Sanjana, who always have been next to me in providing a reason for pursuing my life's purpose. Being a technology nerd, Hemanth contributed in reviewing certain sections of the book for a more appropriate presentation. Sanjana provided her constant energy and motivation.

And importantly, my childhood playmate and one of my best friends, Mr R. G. Rajan, Chartered Accountant, Chennai, who inspired me and guided me through the writing of this book. Many know that Rajan has already authored a book on Auditing.

Finally, I am grateful to Ms Kalpana Shukla, Managing Director, KW Publishers, who invested trust and confidence in my book and offered to publish it. I am also thankful to the entire team of KW Publishers for their efforts and support in making this book reach your hands.

Hyderabad, India
April 15, 2015

Srinivasan M. K.

Preface

Over the last couple of decades. IT and Internet have dramatically changed our life. As more and more consumers and businesses embrace IT and Internet, the business processes and models have to be redefined.

As a result, the professional impact on the accounting and auditing profession has become significant. Accounting and cost accounting have become totally automated, leaving an accounting professional to focus on intellectual work such as business valuation and in establishing newer ways for an enterprise's internal controls etc. So, auditing professionals have been compelled to take a deeper dive into IT and Internet for assessing risks and forming opinions about the fairness of the financial statement prepared and presented by enterprises.

Inevitably, global and regional education-governing bodies have to update the educational curriculum with the latest developments in practice and in the industry.

In this book, I have attempted to capture all the information about business processes, computer hardware, system software, application software, databases, networking (including Telecom network), Internet technologies and emerging technologies. I have ensured that the language is kept as lucid as possible for an easy understanding by a beginner in accounting and by an auditing professional—both of whom will try to understand the applications of IT and Internet in business.

I have kept in mind the various roles of finance professionals while writing this book to ensure that they are able to personally connect the content of this book with their IT applications in their respective work.

- If you are a student of the Public Accounting Course, reading this book will help you to strengthen your understanding of the subject Information Technology, which is the foundation for the Information System Control Audit subject and thus for performing well in your examinations.
- If you are a statutory or internal auditor, the knowledge of concepts and an overview of Information Technology is needed to perform the System audit. This book will provide required the knowledge and understanding.
- If you are a Controller, Finance Executive or a CFO, understanding Information Technology and leveraging the same for business will be of benefit when you evaluate CAPEX investment in IT. This book will provided the required knowledge and understand better the IT and Internet concepts.

Hyderabad, India
April 15, 2015

Srinivasan M. K.

Section 1
IT: Finance Perspective

1.1 Background

Given the fact that Information Technology has invaded all aspects of life at all ages and generations, perspectives differ from individual to individual.

This book attempts to bring together the various aspects of Information Technology from the perspective of a Finance Professional.

The term 'Finance Profession' includes the roles of financial analysts, controllers, CFOs, auditors, financial accountants, cost accountants, management accountants, finance consultants, capital market advisors, investment managers in banks, mutual funds, insurance funds and so on. In addition, irrespective of the level of experience, all finance professionals are included—right from the generation of 'Elders' to the generation 'Millennial'.

I belong to a generation that is at the cusp of the 'Baby Boomers' and 'Generation X' and from a professional point of view, I call my generation as a sandwiched generation—meaning, the generation that has been sandwiched between the traditional way of doing business and the modern business models.

Also, my generation has been lucky to experience the 'Book & Ledger' model of accounting to the 'Click & Close' model of book keeping.

Considering the need for knowledge and skills on Information Technology for accountants, the International Federation for Accountants (www.ifac.org) clearly spelt the need for an exposure on Information Technology for the professional accountants,

> "To help professional accountants worldwide meet ongoing and changing information and rely on information technology challenges, the International Accounting Education Standards Board (IAESB), an independent standard-setting board within the International Federation of Accountants (IFAC), is proposing new guidance outlining the knowledge and skills necessary to prepare professional accountants to perform competently in one or more information technology roles.
> The proposed new International Education Practice Statement, an exposure draft entitled Information Technology for Professional Accountants, provides details of the knowledge and skills required from the professional accountants in the IT environment and to prepare them to use information technology, work in the information technology environment, and rely on information technology...."

This book attempts to provide an exposure to the various technologies relating to Information Management, Internet, Telecommunication and their application in business to significant extent; if not to the full extent as recommended by the IFAC.

1.2 Accounting Processes History: 'Book & Ledger' to 'Click & Close'

The golden rule of accounting and principles behind debits and credits have not changed since the ancient days, but the way we maintain books of accounts have dramatically changed. Though the current generation is using sophisticated computers to perform accounting functions, there is no certainty whether the current-generation finance professionals are aware regarding how the financial books were maintained in the olden days through subsidiary books for each function such as Purchase, Purchase Return, Bills Payable, Sales, Sales Return, Bills Receivable, Cash, Bank, Journal Entry Book (for passing financial period end adjustments). All the books were manually updated, manually totalled (with the help of mechanical calculators/electronic calculators in the later times), literally tallied and reported. The General Ledger would have the periodic summary of all these subsidiary books posted and accounts finalised.

The principles of subsidiary books are still applied in the current-day ERP (such as SAP) through a design called 'Sub-ledger'.

In the1950s and 1960s, Mainframe computers were deployed by major corporations in the banking and insurance sectors. This involved data entry through offline data entry machines. The medium of data were punched cards and 8" floppy diskettes.

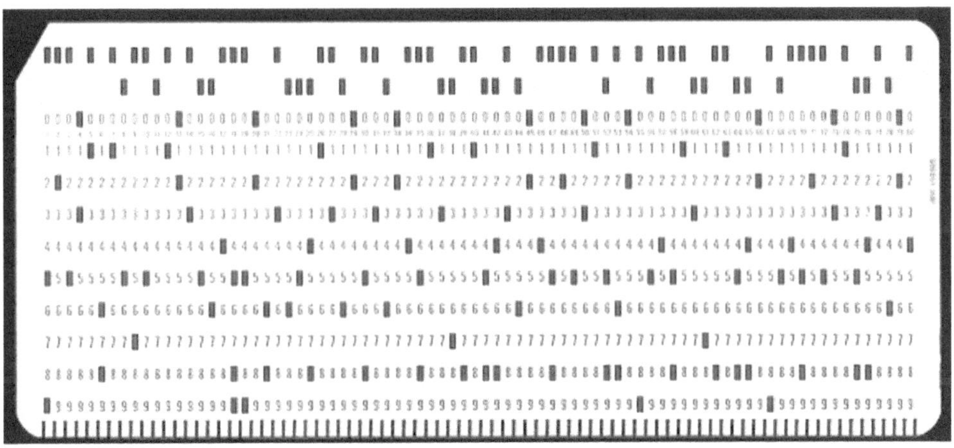

Diagram 1.1 Punched Card

Diagram 1.2 Floppy diskettes of Various Sizes

Mainframe computers were very expensive to maintain and were affordable only by very few large corporations. Medium-size and small-sized enterprises continued to have manual book-keeping.

The 1970s were the era of 'Personal Computers' (PC); A computer system that was made affordable for the medium and small businesses. Almost every enterprise realised the value of having a computer. The businesses were run in the same old fashion with most of the enterprises having a functional departmental structure. Either the small EDP (Electronic Data Processing) Department was having a bunch of PCs and performing computing operations for different functions as a shared service or each department had its own personal computers applied to its business operations. During this era, most of the manual functions were automated. Depending upon the investment levels in IT, each enterprise had its own automated system. Custom-built financial accounting, payroll, MRP I, MRP II and inventory management were popular systems in those days.

Functional automation was very popular and financial accounting books were automated by various software enterprises. The most popular one in those days was the Oracle Financial.

In India, Tally was one of the first financial accounting softwares.

In the 1980s and early 1990s, networking technologies were invented and commercialised. The enterprises embraced client-server architecture. Novell Netware was the most popular one which was compatible for the MS-DOS PCs. Prior to Novell Netware, various variants of UNIX were existing. More networking technologies such as Windows NT, Linux, etc. came later.

Then came the Enterprise Resource Planning (ERP) and around the same time Michael Hammer propounded the Business Process Reengineering (BPR). At one point in time there were hundreds of ERPs worldwide, though the popular ones were SAP and Oracle.

Technology further enabled Data Warehousing, which in turned enabled Business Intelligence. Internet further accelerated and facilitated the possession of a uniform system and processes with a single instance of an ERP for an enterprise globally.

The convergence of Information Technology, Communication and Entertainment enabled the accessing of enterprise applications through mobile devices. Alongside the technological advancement, new business models were developed such as e-Commerce, Social Networking, Customer Self-Service models in Banking, Insurance, Airline services, Travel & Transportation, Hospitality sectors, etc.

As against the conventional way of recording the accounting entries on a receipt of documentary evidence for the transaction, in the current era the accounting entries are posted in books as and when transactions were triggered. For example, during the olden days, if you wanted to withdraw cash from a bank, you would visit the bank and submit a written requisition in the form of a withdrawal slip or a cheque and upon submission, the cash would be disbursed after verifying the signature to identify the account holder and the payee. The cheque or withdrawal slip therefore becomes the documentary evidence for transaction and the individual account balance is debited and cash is credited and in the General Ledger the deposit account is debited and the cash account is credited.

Nowadays, we use an ATM card to withdraw cash; identity is managed with the help of a physical card along with a Personal Identification Number (PIN). As the ATM machine disburses the cash, the accounting entries are automatically posted in the bank's books of accounts. In the same way, accounting entries are automatically posted in the company's books based on the transactions triggered by the various business personnel.

It is a paradox that the CFO is responsible for the financial accounting records, but except period-ending accounts' close-related transactions and certain financial entry transactions such as depreciation posting, remaining transactions—primarily revenue and expenses—are automatically posted in the books of accounts based on transactions triggered by someone outside the Finance and Accounts department. Strange—but true.

Added to these complexities, the Internet contributed towards an accelerated technological advancement. Global corporations nowadays transmit their commercial documents, payments to their business partners through internet protocol and those transaction records are captured by the business partner's information system which in turn triggers accounting entries. In a nutshell, entities outside one's own organisation trigger transactions and consequential accounting entries in the books of accounts.

1.3 Volatility, Uncertainty, Complexity and Ambiguity (VUCA)

The VUCA explains the current ways of modern business. Every business in every industry is confronted with VUCA.

VUCA is caused by

1. Technological advancement: The example of Nokia—once a leader—now no longer in the mobile-phone business.
2. New business model: The e-Commerce business model is fast replacing the brick-and-mortar models.
3. Economic changes: The European Union is confronted with economic challenges as are all the members of the EU are confronted with economic challenges.
4. Power shift: Eastern countries such as China, India are developing economically and are thus beginning to possess bargaining power. Companies from these countries are becoming global corporations.
5. Political impact: Besides competition between enterprises, nowadays the government under different political philosophy competes to garner foreign investments in each nation.
6. Terrorism and money-laundering are newer risks (if not a threat) in conducting business in certain geographical areas.
7. Generational changes in terms of consumption, living outlook about business and consequential social changes such as the collapse of the monetary system such as 'no value' derivatives through securitisation.

The list can go on.

All of the above and more of similar factors cause VUCA. A business can no longer planned for long term—say, 3 or 5 years.

1.4 Impacts of Various Changes on Finance Profession

A combined effect of the technological developments in Information Technology, Internet Technologies and VUCA have caused a significant impact on the Finance and Audit profession.

Finance and accounting professionals are now completely relieved of maintaining books of accounts and this has been delegated to computers. This has resulted in the finance professionals being left with more time for analytical work (on a lighter side, they are also left with the time for finding software bugs in the financial systems!)

Besides the change in the activity of maintaining books of accounts, there were several other substantial changes in the professional activities:

Auditing: The auditing approach has changed from 'document/voucher verification' to 'system study'.

Internal Control: Evaluation of internal control has shifted from evaluating 'roles and responsibility of individuals' to a 'role-based authorisation in system to trigger transactions' in a computer environment.

Validating Data: The evaluation method of 'independently verifying with external sources such as debtors, creditors, banks' has changed to the evaluation of 'data veracity and integrity of the entire system'.

Financial Exposure: In the earlier days, there was a 'reviewing of risks/events and their impact on the financial liability exposure'. Nowadays, there is an additional need for reviewing the information system for a lack of data security and privacy to assess possible liability and in providing for the same.

Assessing Enterprise Strategy and Management: This is done for assessing the forward-looking statement made by the executive management. In addition, there is a strong need for assessing the data governance and governance of the information system.

Valuation: Unlike conventional approaches to valuation, current technologies such as Internet offer sources of information for the valuation of assets. Still, valuation continues to be a challenge.

Assessment of True and Fair/Correct State: Auditors are compelled to look at factors beyond numbers to assess true and fair/correct state of business—particularly the IT set up in an enterprise.

Liability under SOX and SAS70: Post Enron and WorldCom events, regulatory compliance needs are expanded to cover enterprise management executives such as the CEO, CFO and the auditors are responsible to the investors/shareholders.

CFOs and Enterprise Financial Management: The roles of the CFO compel them to get deeper into the enterprise operations and technical aspects of the business for deciding on resource allocation for various enterprise activities/operations for delivering the optimal business results.

CAPEX Decision to Seize Technology Deployment in Business for Revenue Maximisation or/and Cost Reduction: This demands the CFOs to understand the technological advancement happening around and its relevance for the application in one's own business. More education on technological (particularly Information Technology) advancement is needed for the CFOs and other finance professionals.

Real Time Enterprise Performance Management: With the support of advanced technological development, enterprises are now enabled to monitor Enterprise Performance Management (EPM) in real time. This calls for the CFO to design the EPM for continuous monitoring.

Treasury Operations: Treasury operations, particularly investment management through capital market operations, demand deployment of a sophisticated software tool for timely decision-making to minimise loss/stop loss. This demands the CFOs and finance professional to understand the way in which these softwares are functioning space.

The list is unlimited and can go on.

The crux is that unlike in the past, the expectations and demands from finance, accounting and audit professionals are that they gain a knowledge beyond finance and accounting—particularly of Information Technology.

This book attempts to provide knowledge on the fundamentals of Information Technology (IT) such as Hardware, System Software, Application Software, Databases, Network, Internet Technologies, Telecommunication Applications and how the application of IT has compelled the enterprises to redefine the business processes. Admittedly, some of the topics might have been learned by current-generation finance professionals as part of their curriculum during their previous education. However, this book attempts to provide a holistic view of Information Technology and other technologies such as the Internet. In addition, the book also covers certain other advanced and emerging technologies.

1.5 What not to Expect from this Book

This book will not tell you

1. How to audit a business in the IT environment with the best practices.
2. How to govern the IT environment in a business enterprise.
3. What to do or what not to do relating to investment in the IT set-up.
4. How to take decisions with respect to CAPEX investment in IT.
5. How to configure an ERP system.
6. Which software to be chosen.
7. Whether to choose a private cloud or public cloud or hybrid cloud.

Also, questions similar to the above.

The main reasons for not covering the above are

1. This book provides a fair amount of knowledge on the various technologies available to choose from, when it comes to adopting IT and related technologies and leaving the decisions to the individual finance professional.
2. There is no 'one-size-for-all' kind of solution available in IT and related technologies.
3. Finally, the decisions depend on enterprise policies, risk-taking abilities, investment capability and expense affordability.

Finally, the times have dramatically changed and changes are happening faster than in the past. Unlike in the past, in the current era of fast-changing times, the professions demand a continuous learning. Finance, accounting and auditing professions are not exceptions to this.

Section 2
Business Process and Automation

2.1 Background

Information Technology and automation have offered several distinctive advantages for business. One such advantage was the application of computers to various business functions. In the latter part of the last century, several software applications automated different business functions. These applications just brought automation to the then existing processes, but did not truly transform the processes.

Then, each of the business functions were operating in a silo, as it called for specialized skills. For example, in order to work in an accounting department, a person should have the knowledge of double-entry book keeping; similarly, for other functions also. With the result, each of these functions—Finance-Accounting-Sales-Marketing-Procurement-Inventory Management-Production Planning-Manufacturing-Transportation—were working in silos.

These functions have been automated through software applications and the digitisation of data.

In the 1990s, these types of functional automation were challenged and specific end-to-end business processes that enable value to business and customers were evolved. Thus, the birth of business processes.

2.2 Business Process Management

What is Business Process Management (BPM)?

BPM is a set of processes that evaluates and assesses business processes set up in an enterprise in the light of

1. Whether the processes are reducing costs
2. Whether the processes are creating value for customer

Business process management is not a stand-alone programme, but an integral part of enterprise strategy. BPM enables the enterprise to align its process to the enterprise goals. BPM enables the institutionalisation of a process-oriented culture in an enterprise. Tracking and monitoring the performance of processes is part of BPM activities. A cross-functional team is involved in managing the business processes. BPM is achieved through an organisational transformation process called 'Business Process Reengineering—often called as BPR—'Total Quality Management' (TQM) and 'Six Sigma'. It is deployed with a combination of processes, methods and tools.

Transformational Processes

There are several transformational processes that are invented from time to time for enhancing the business effectiveness and profitability of the enterprises. Here are three transformational programmes that help the business process to improve:

Business Process Reengineering (BPR)

BPR is the fundamental rethinking and radical redesign of processes to achieve a dramatic improvement in the critical contemporary measures such as cost, quality, service and speed.

Total Quality Management (TQM)

TQM is an organisation-wide effort to set a permanent culture and climate for quality management and continuous process improvement.

Six Sigma

A set of techniques for process improvement, it seeks to improve quality by preventing the defects and by minimising variability in the manufacturing or business process.

What is a Process?

In order to understand and take a deep dive into Business Process Management, it is necessary to understand the meaning of a process.

A process is a sequence of events using certain inputs to generate outputs. Also, a process is a well-coordinated rhythmic flow of activities performed by people and machines involving different functions for creating business value to achieve a business goal.

Functional Process vis-à-vis the Business Process

In the pre-ERP days, the processes existing in businesses were more function-specific. As explained earlier, there were accounting processes in the accounting function, sales processes in the sales function, procurement processes in the purchase function and so as in the production planning and manufacturing functions also. Invariably, these processes were triggered and ended within the function itself.

In the current era, as against functional processes, we have end-to-end business processes. The business processes can be triggered either within the organisation or from outside. Business processes will cut across functions and people from different functions will be part of the process team.

Here are the differences between a functional process and a business process:

Attribute	Functional Process	Business Process
Organisational unit	Functional department	Process team
Leadership	Functional leader	Process leader/Owner
Advantages	Focus is on functional specialisation. Easier to execute work distribution as all members in the functional department have similar skills. Performance evaluation is based on how management directions are executed.	Processes are laid based on customer and market needs. Enable collaboration and communication among teams of different functions. Performance evaluation is based on the achieving process goals.
Disadvantages	Lack of collaboration and communication between functional teams. Lack of effective coordination between teams. Sub-optimised business process as an end-to-end process is lacking.	Redundancy of functional skills. Performance evaluation not in line with functional expertise. Complexities of end-to-end processes.
Business value	Value for cost leadership strategy enterprises	Value for product differentiation and disruptive innovation strategy enterprises

Process Management

Defining Process

The first step in process management is to define the business process. The definition of business process is derived from the strategy of the enterprise and the policies and procedures associated with the strategy.

Organisational Set-Up

The second step is to set up the organisation with people to execute the defined processes. People are going to decide as to when to process which step in the process under what circumstances.

Compliance Testing

The third step is to periodically verify whether the set processes are being followed by the organisation. The objectives of verifications are twofold: (1) to ensure that the set processes are being followed by the organisation; identify if there is any process violation (2) to verify what are the exceptions thrown up by the business transactions and to understand the event to refine the processes.

Performance Measurement

The next step is measuring the performance of the process and the business. Measurement of performance is very critical from the point of view of the strategic intent of the enterprise. The set process is expected to make the business operations smooth and ultimately the enterprise has to realise the strategic objectives from the process operation. From this perspective, the performance has to be compared with the targeted performance level.

Definition of BPM

It is the achievement of an organisation's objectives through the improvement, management and control of the essential business processes.

Every word in the definition is very important in the defining and understanding of BPM. In the context of an ***organisation,*** it is to ***achieve*** the business ***objective,*** to ***manage*** *and* ***control*** the ***business processes*** and also to ***continuously improve.***

From the definition it will be very clear that the processes are to achieve the enterprise's business objectives and business objectives are derived from strategies. Essentially, strategies and business processes are linked.

Principles of BPM

Business Processes are Assets of the Enterprise

Every business defines its unique processes to create value for the enterprise itself and for the customer. Therefore, processes are assets of the enterprise. In fact, when any process is defined and the same has been configured in a packaged software or coded with the help of a developing custom software, the said processes set up in the software become a proprietary asset of the enterprise.

Creates Value for the Customer

Through the process definition and implementation, the overall cost of doing business is reduced and thus the customer is benefited with a better value for the money paid or reduction in price. For example, when a cab operator launches a smartphone-based application for booking cabs and payment, the enterprise encourages the self-service method for customers. The costs towards call-centre operations are reduced. In this scenario, the customer is benefited with the flexibility of self-service, at one's own convenience and the cost of service also may be reduced by the cab-service provider.

Continuous Improvement of the Process

BPM includes compliance testing of the process and one of the objectives of compliance testing is to ensure the continuous improvement of the process. Continuous improvement is needed for various reasons such as leveraging technology advancement or new regulatory compliance. Therefore, it is one of the principles of BPM.

BPM Best Practices

In order to derive maximum optimal benefits out of implementing BPM in an enterprise, the following best practices are recommended:

Strive for Process-Oriented Organisational Structure

By adopting BPM, an enterprise commits to a process-oriented organisation and a process-oriented way of doing business. While a functional structure may still exist, a process-oriented structure and team are critical for success.

Appoint Process Owners

As part of a process-oriented structure, there will be a strong need for naming the process owners or process leaders. The process owner must be entrusted with the responsibility of not only the effective and efficient functioning of the named process but also for its continuous improvement.

Senior Management Commitment to Drive BPM

No initiative in any enterprise can succeed and bring the fullest benefit unless the commitment of the senior management to the initiative is experienced. BPM is not an exception to that.

Bottom-Up Approach for Process Improvements

Process improvement exercises have to be started from the bottom. Listening to people who run the processes is important. Enterprises that are successful in BPM implementation have encouraged every level of people in the organisation to critically evaluate and engage in its continuous improvement.

Place IT to Monitor, Control, Analyse and Improve Processes

Leveraging Information Technology for accelerated benefit of BPM is very common these days. Once the end-to-end enterprise processes have been defined, automating with the help of deploying the IT suite of BPM tools will immensely benefit any enterprise. Besides, operating the process in an automated fashion. IT can help the enterprise to gather valuable data on the functioning of the process itself. Those data will help the enterprise in analysing and improving the process.

Work in Collaboration with All Business Partners

While it is a prerequisite that all the functional people within the enterprise come together and work towards process definition the operate the process and engage in continuous improvement, currently the global economy and inter-enterprise cooperation and collaboration with the external business partners for the success of the BPM is also essential. External business partners include marketing partners, sales channel, end customers, vendors, service providers, etc.

Train the Workforce

Another critical success factor for BPM deployment in any business is training the workforce. The primary reason for training the workforce is to familiarise it with the following:

- Why the BPM is being implemented?
- What is the strategic intent behind BPM implementation?
- Overview of the end-to-end enterprise processes.
- Information on intersection and interaction of processes and their dependencies.
- Regulatory compliance needs of the process.
- "Do's and don'ts" for a given process.

The above list is only illustrative and not exhaustive. Once the workforce is trained with "what we are doing", "why we are doing" and "how to do", it is empowered and voluntary commitment is secured. Workforce understanding, empowerment and voluntary commitment are the most critical success factors for the success of BPM deployment.

Align Employee Incentives/Bonuses to Business Process Performance

Enterprises that are successful in BPM deployment have successfully linked the employee incentives and bonuses to the performance of the process and ultimately to the performance of the business. It is a very good approach to link employees' performance-related remuneration to process performance.

Use Incremental and Radical Methodology for Process Improvement

In order to derive maximum benefits out of BPM, successful enterprises have resorted to using both an incremental methodology and also a radical methodology in process improvement.

Invariably, when the BPM is first set up, a radical approach may be needed. Business Process Reengineering (BPR) is a radical approach methodology. BPR is radical as it is a transformational exercise for an enterprise as enterprise processes, organisational set-up and way of conducting business—all aspects of an enterprise get changed.

Once the BPM is deployed, then the enterprises resort to continuous improvement as a part of it. Continuous improvement is not radical, but will bring small benefits every time.

All successful deployments of BPM, have implemented it using both the radical and incremental methodologies.

2.3 Overview of Business Process and Process Flow

Business process is defined as "A sequence of work steps performed to produce the desired results".

The entire life-cycle of a process with the sequence of steps is called "process flow". A process flow may contain different process steps within a flow. The accounting process flow is also known as Record to Report.

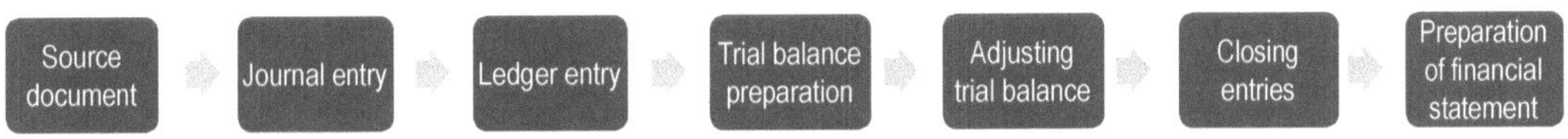

Diagram 2.1 Accounting Process Flow

Each of the process flow steps is a business process—For example, a source document will have a different set of rules attached to it: for customer transactions sales invoice will be the source document, for a vendor transaction purchase invoice will be the source document and for financial entries for accrual, it will be a journal voucher and there are a set of rules for processing each of the document. Similarly, a journal entry will be based on the transaction nature determining whether it is a personal, real or nominal account.

Examples of process flows are

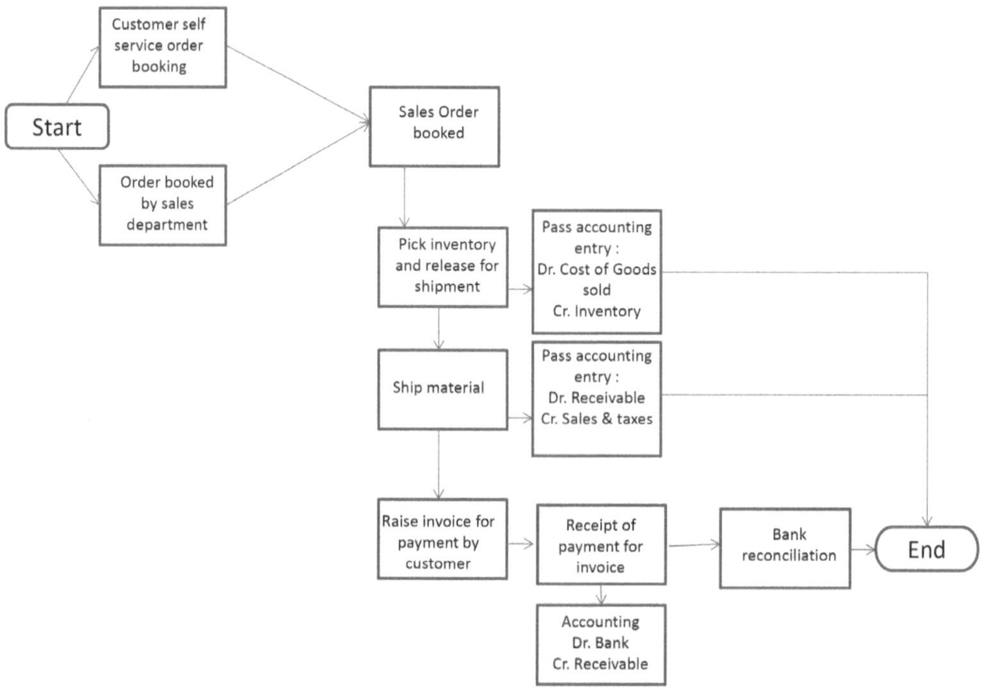

Diagram 2.2 Order to Cash Process Flow

On reviewing the above process flow chart one can observe the following:

1. The process is end to end
2. It is independent of any specific function
3. It cuts across different functions as explained below:

Process Step (Order to Cash)	**Function Affected**
Order booked by sales department	Sales function
Sales order created	Sales function
Pick inventory and release for shipment	Inventory management function
Ship material	Shipping department function
Raise invoice for payment by customer	Sales or shipping function
Pass accounting entries for goods issued for shipping	Accounting function
Pass accounting entries for sales revenue recognition	Accounting function
Receipt of money, accounting and bank reconciliation	Finance function

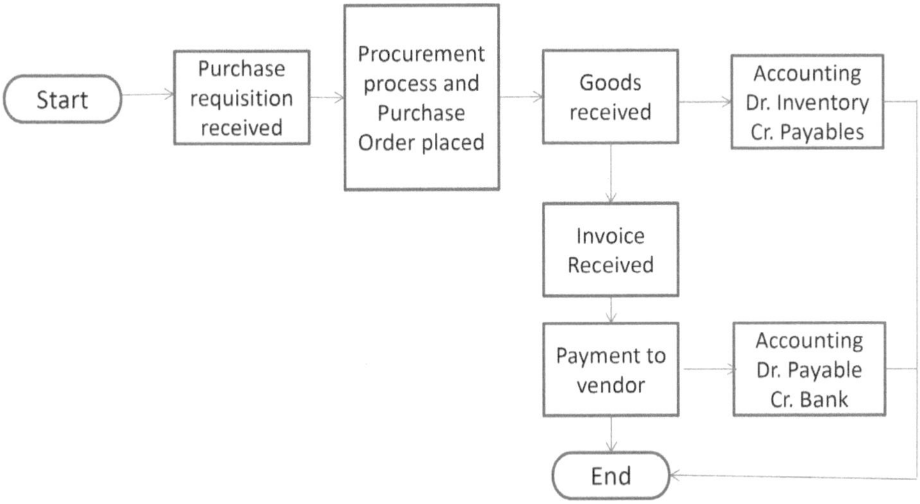

Diagram 2.3 Procure to Pay Process Flow

Process Step (Procure to Pay)	**Function Affected**
Purchase requisition receipt and placing order	Procurement function
Goods receipt	Inventory management function
Invoice receipt	Accounting function
Accounting payable to vendor	Accounting function
Accounting entries for inventory	Accounting function
Payment to vendor	Finance function

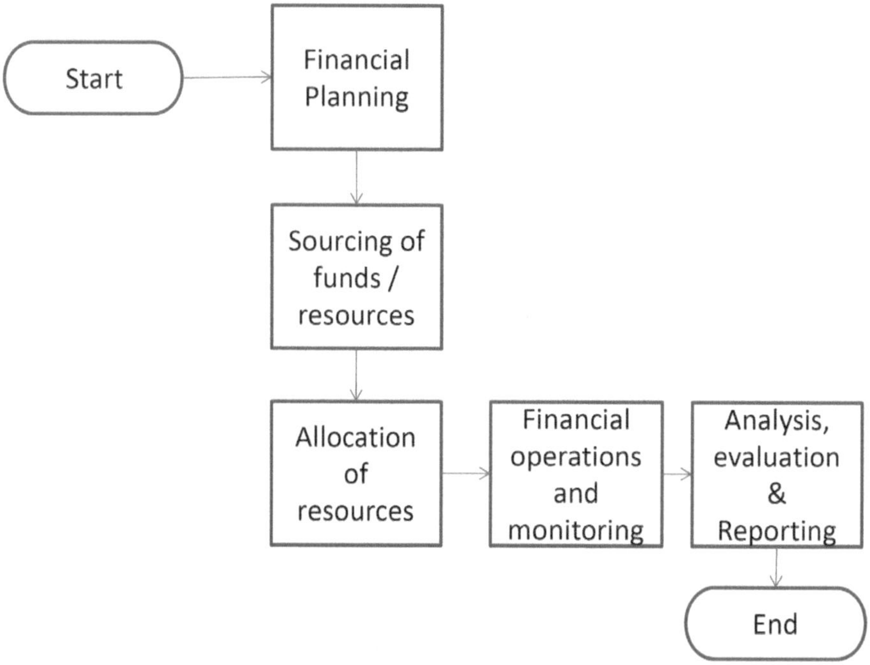

Diagram 2.4 Financial Management Process Flow

The above financial management process is simple and self-explanatory that explains the sourcing of funds, application of funds, financial operations such as treasury, etc., monitoring, reporting, analysis and evaluation.

2.4 Classification of Business Processes

Enterprise Strategy, Goals, Organisational Processes and Operational Processes

Enterprise strategy triggers lower-level structure, processes and controls. In other words, the lower-level operational processes and middle-level management processes are designed and deployed to enable not only the achievement of goals but also to measure the goals and in turn realise the outcome as envisaged in the strategy.

Diagram 2.5 Business Processes at the Different Levels in the Organisation

In the above pyramid, the downward arrows indicate design and direction. The upward arrow indicate the reporting of the process outcome, the measuring of metrics and the realisation of the strategic outcome.

Organisational Business Processes

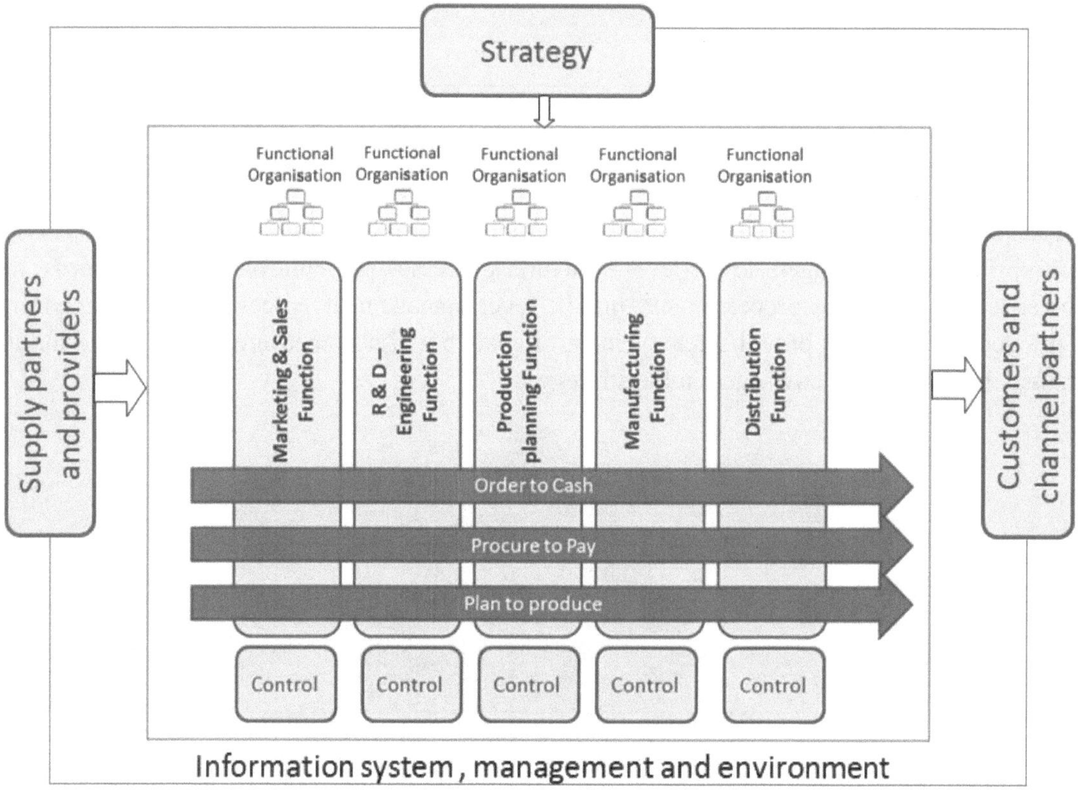

Diagram 2.6 Organisational-Level Business Processes

In the above diagram, the entire business process ecosystem is explained: Strategy at the highest level, below that the functional management process that is designed and set up and operational business processes cutting across the functional organisations. On the periphery, supply partners and providers exist and on the other side of the periphery we have customers and channel partners. All these are managed through the information system.

Operational Processes

The operational processes are a set of activities that are executed to convert the raw materials into finished products and deliver the product value to the customer. The operational processes are:

Order to Cash

This process includes securing a sales order from the customer, shipping the product and realising the money value from the customer. This process cuts across the marketing and sales, inventory management, manufacturing and finance functions of the enterprise.

Procure to Pay

This process enables procurement activities such as search or supply partner; select; place order; receive goods and effect payment to the supplier. This process cuts across the procurement, inventory management and finance functions of the enterprise.

Hire to Retire

This process relates to the Human Capital Management of an enterprise. The entire process relating to talent search, hiring, induction, orientation, development, progression, compensation and retirement processes of the Human Capital Management function are mapped; as the human capital is common for all functions, this process cuts across all functions.

Record to Report

As each and every transaction in a business is carried out and if it is involving any monetary value, it has to be recorded in books of accounts. At the period end, the recorded transactions have to be reported in a suitable way based on whether it is financial reporting or management reporting. Again, as the business transactions are transacted across different functions, the impact on the monetary value has to be recorded in books of accounts; this process cut across all functions.

Plan to Produce

The aforesaid business processes are common to most of the business entities. In case of an enterprise engaged in manufacturing, this is an additional process that has to be mapped. The primary function of this process is to plan the manufacturing resources and the material requirement. In addition, this process enables production scheduling also.

2.5 BPM Implementation

BPM Implementation: Factors to Consider

In any BPM implementation in an enterprise, the following factors have to be considered:

Scope

The scope has to be clearly defined. The scope could be selecting a single business process as a pilot run or taking up all the processes in the business simultaneously as a 'Big Bang' approach. The essence is a clear definition of the scope of BPM implementation.

Goals

The goals to be achieved through the BPM implementation are to be clearly stated. It could be a simple 'Process Improvement' or 'Optimisations' or a radical 'Business Process Reengineering'. In each case, the goals have to be clearly stated.

Method Selection

There are several tools available to implement BPM such as BPR, benchmarking, Six Sigma, etc. An enterprise may choose to go with its own developed informal method. The crux is that it is important to choose and decide on the implementation methodology.

Human Resource Needs

In order to implement a successful BPM implementation, an enterprise needs the necessary skills and the human resources possessing those requisite skills. The skills required typically are: consultants who have expertise in BPM implementation methodology; whether a software product is selected for implementation and technical and functional consultants who have requisite skills in the software. Ideally, successful BPM implementations have dedicated human resources from the enterprise with the required functional experience.

Tools

As BPM implementation involves a good amount of brainstorming discussions and extensive documentation of processes, it is necessary to have necessary tools that facilitate discussions and documentation. Also, at different stages, there will be a need to train people and therefore, tools for conducting training will also be needed.

Resource Investment

An enterprise aspiring to implement BPM needs to make a necessary investment for acquiring the necessary resources for implementing it. Investment in human resources, time, acquisition of skills, tools, software, investment for training are needed.

Sponsorship and Commitment of the Executive Management

Like any other enterprise-level initiative, the implementation of BPM needs sponsorship and commitment of the executive management. Sponsorship activities include a formal announcement of the project, naming a project manager, committing investment, conducting a periodical review, facilitating timely critical decisions and sending directives to all levels of the organisation to enable every employee to understand how mission-critical the BPM implementation is.

Need for BPM

An enterprise needs BPM for the following reasons:

1. BPM enables long-term future positioning in the competitive environment—of late, having BPM may not help with a competitive advantage, but not having BPM can hurt the enterprise in the market.
2. Helps the enterprise in achieving cost effectiveness and effective customer service.
3. BPM lays the foundation for a continuous process improvement.
4. BPM facilitates business-intelligence reporting and the enterprise is in a position to gain a view of product profitability and customer profitability and to evolve suitable business strategies.
5. A radically reengineered business process helps the enterprise to derive a quantum jump in cost savings.
6. BPM enables teamwork and empowers the employees to go beyond one's own functional boundary for teaming. This facilitates cross-functional teaming and hierarchical reporting.
7. BPM enables the creation of new roles such as process leaders for some of the functional managers and helps in empowering all employees in the enterprise.

BPM Implementation Success

The BPM implementation's success significantly depends upon the following:

Design

Designing and developing the business processes and enterprise architecture are critical to the success of BPM Implementation.

Value Add

Ensuring that the designed processes aligned to business strategy, compliance requirements and control become integral parts of the BPM critical to the success of its implementation.

Integration

Integration of IT systems with business processes and integrating teams—external consultants with internal teams—are again critical success factors for BPM implementation.

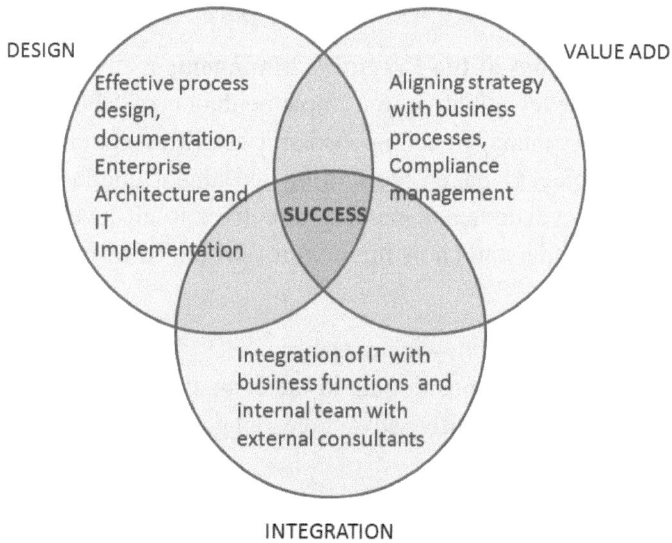

Diagram 2.7 Success Factors for BPM Implementation

Functional Automation and BPM Implementation

In the 1970s and 1980s, when the personal computer was brought to the market, every business leveraged its application for business process automation.

The only advantage of the functional process automation was to digitise the record-keeping of the function and to leverage the computer capability for automation. Otherwise, there was no process improvement or synergy among all the derived processes.

Functional automation does not involve any radical thinking or reengineering of the processes. All the deficiencies existing in the manual functional process continue to exist in the automated functional process also, except speed in processing.

Functional automation created islands of automated system operating in silos, with lots of challenges such as redundancy of data, lack of ability to flow data from one system to another, etc.

Often, functional automation is mistaken for business process automation.

The diagram below explains functional process automation:

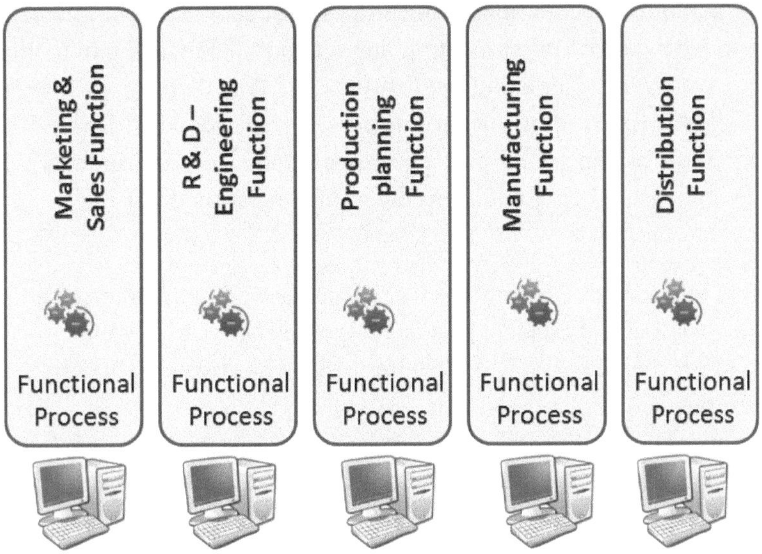

Diagram 2.8 Functional Process and Automation

On the other hand, Business Process Automation is involving radical thinking and reengineering or redesigning of the processes. The process thinking is an end-to-end activity, without thinking about functional boundaries. The process may cut across several functions when it flows and activities are performed.

The diagram depicted below represents the Business Process Automation:

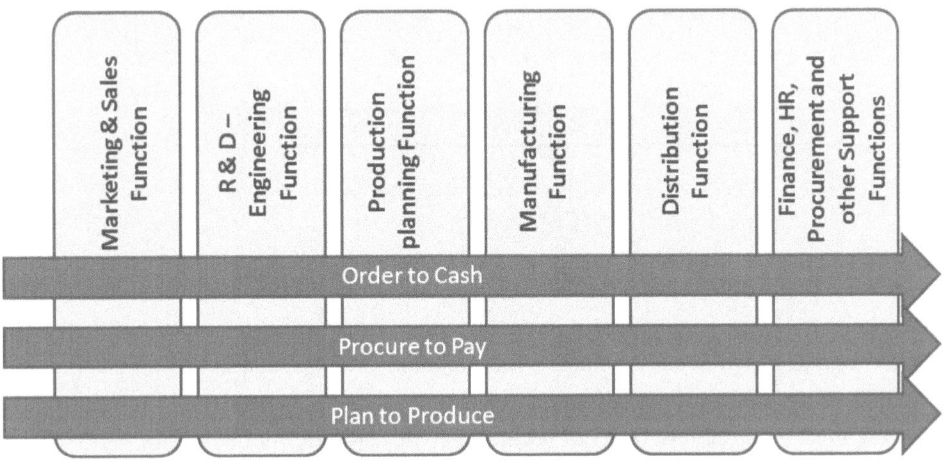

Diagram 2.9 Business Process Automation

BPM Technology

BPM technology solution is a computer application suite deployed for implementing BPM. This is implemented independent of other existing applications. The BPM suite, once implemented, can be interfaced with various existing applications. BPM technology manages the automated workflow and also the functioning of various existing applications. The BPM application is useful for tracking performance and also ensures compliance requirements when configured. Enterprises have immensely benefitted from the implementation of BPM.

Value Chain

Value chain refers to a series of separate activities performed in an enterprise to create value to the customer. This value chain concept was developed by Michael Porter.

Value chain offers a process view of an enterprise. It has two sets of activities:

1. Primary activities
2. Support activities

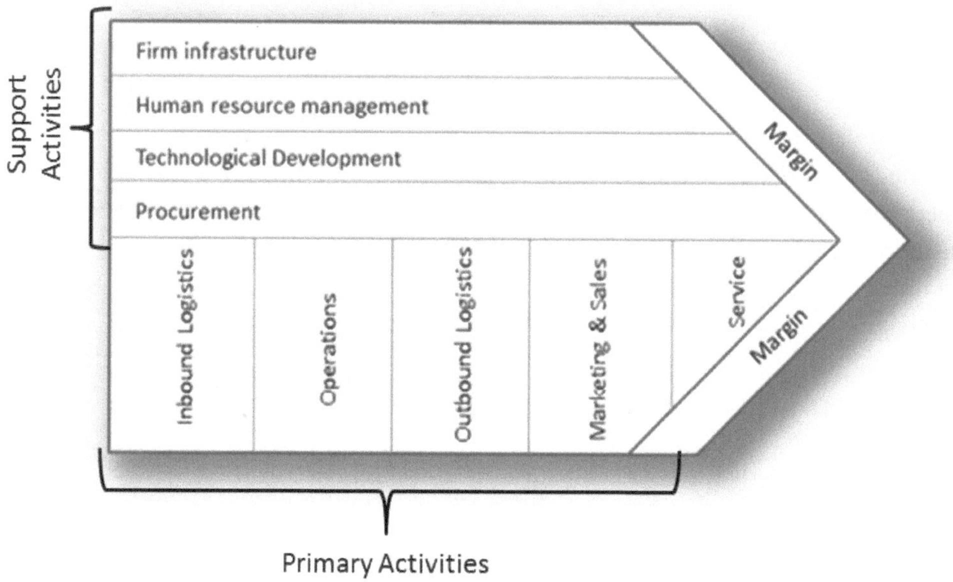

Diagram 2.10 Value Chain

Value-chain analysis is very useful in determining what activities of an enterprise add value to the customer and how they form part of the business process.

Six major value-chain processes are as follows:

1. Research and Development (part of technology development above)
2. Design of products, services or processes (part of operations above)
3. Production/Manufacturing (part of operations above)
4. Marketing and sales

5. Distribution (part of outbound logistics above)
6. Customer service (part of service above)

BPA Advantages

- It helps to save costs by cutting down on the unwanted activities in the process.
- Helps the enterprise to compete well in the market as the BPM helps to create value for customers.
- A well-defined business process along with automation helps in a speedier customer service.

Risks

- When unwanted activities in a business process are trimmed and automated, it may result in a loss of jobs for several employees.
- Unless a well-thought reengineered business process is designed, enterprises may end up automating the functional processes without much synergy. This may result in a waste of investment.

2.6 Accounting System Automation

An accounting system in an enterprise exists for capturing the record of all accounting information as a result of the transactions that take place and is expressed in monetary value. An accounting system is interchangeably called as Accounting Information System (or AIS).

Functions of Accounting Information System

1. **Collection and Storage of Data:** As the business events trigger transactions involving any monetary value, collection and storage of all the monetary value are the first function of the AIS.
2. **Record all Transactions:** Upon their collection, the data needs to be recorded with an appropriate accounting treatment as an accounting journal entry. An AIS is expected is to record data in the form of revenue, expense, asset or liability and enter it in the books of accounts.
3. **Safeguarding of Assets** is another important function of the AIS and this is achieved through the maintenance of the records of all accounting transactions through extensive documentation and appropriate segregation of duties. Besides physical assets, information generated in an enterprise is also considered as a valuable asset.

Reporting is another function of the AIS, though considered as an incidental function. The AIS is expected to deliver reports for various purposes: financial statements for shareholder reporting, management reports for management information and decision-making and statutory reports for compliance needs.

Processing Cycles in an Accounting BPM

BPM processing cycles for accounts are as follows:

- Finance Cycle
- Expenditure Cycle
- Human Resources Cycle
- Conversion Cycle
- Revenue Cycle

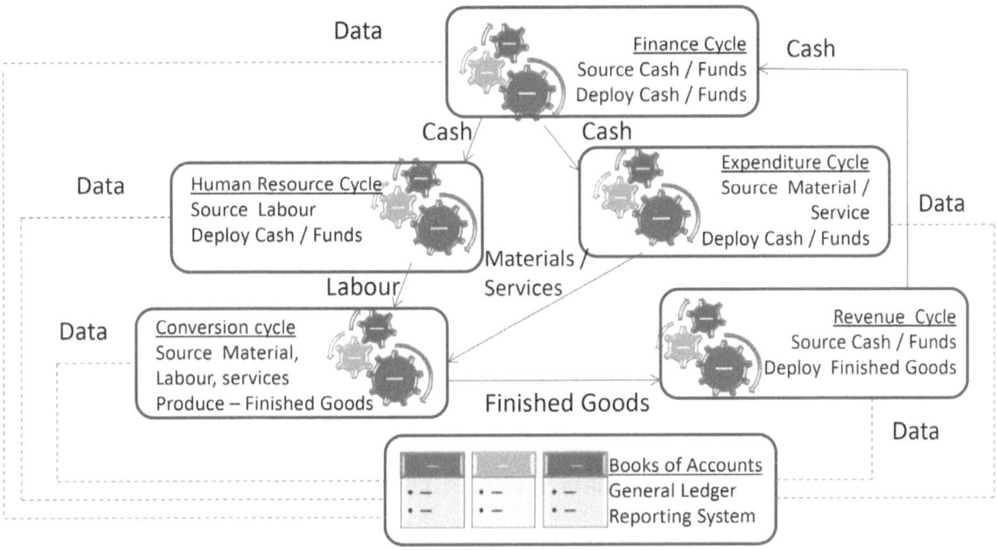

Diagram 2.11 Processing Cycles in an Accounting BPM

Finance Cycle: In this cycle, the process records the sourcing of funds from the sales proceeds of finished goods (and funds raised from external sources through capital and money-market operations). This cycle deploys cash and funds to the human resource cycle and expenditure cycle.

Human Resource Cycle: This cycle mobilises cash resources from the finance cycle, mobilises the labour deploying cash and makes labour available to the conversion cycle.

Expenditure Cycle: In this cycle, monetary resources are made available by the finance cycle and this cycle procures and deploys material and services to the conversion cycle, using the monetary resources made available by the finance cycle.

Conversion Cycle gets material and services from the expenditure cycle and labour from the human resource cycle and produces finished goods and makes them available to the revenue cycle.

Revenue Cycle delivers finished goods to the customer and generates cash/funds, which are made available to the finance cycle.

The diagram given below is the subsystem of the main system explained above. The subsystem consists of three critical cycles: Expenditure Cycle, Conversion Cycle and Revenue Cycle.

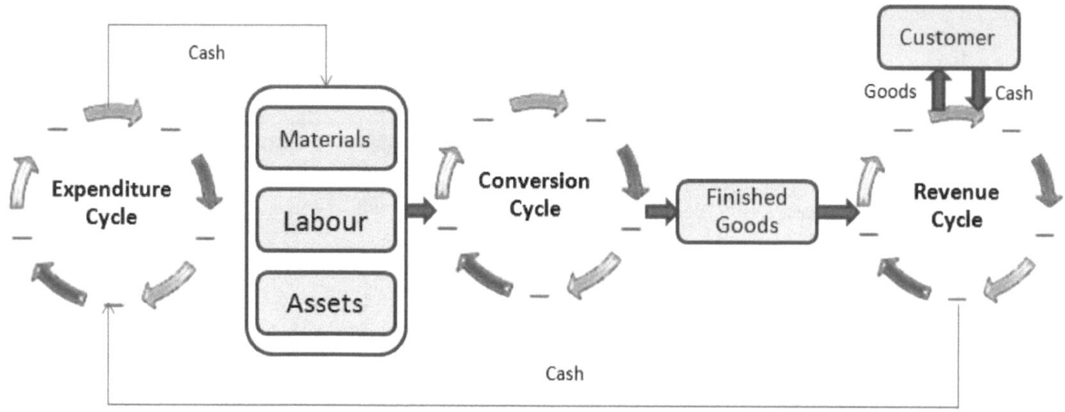

Diagram 2.12 Expenditure, Conversion and Revenue Cycles

Source Documents

A transaction in any cycle can occur only when a business transaction event is triggered. When a business transaction is triggered, the details of the transactions are captured in a document as input for data processing. These documents are called source documents.

A source document is critical from two perspectives:

1. Source documents are a must to have firstly, in order to record the transaction in an automated BPM environment and ultimately into the books of accounts.

Diagram 2.13 Transaction: Source Document to Financial Statements

2. Secondly, when the audit of accounts is being conducted, the financial statement numbers should help the tracing back to the source documents. This information is generally traced from the audit trial.

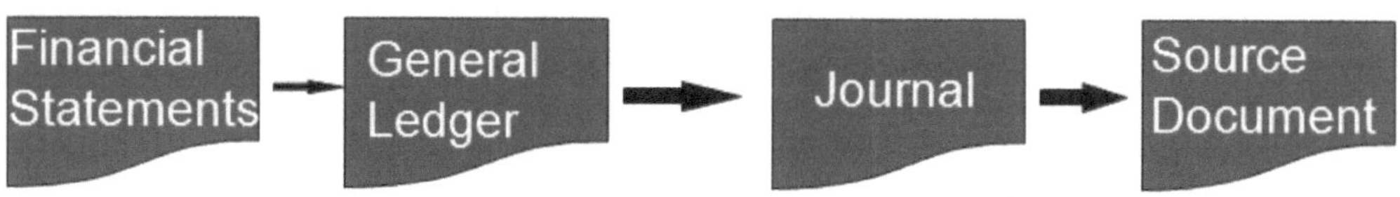

Diagram 2.14 Audit: Financial Document to Source Document

Source Documents for Revenue Cycle

Source Document	**Representing Transaction**
Sales Order	Evidence of customer order
Delivery Note	Evidence of delivery of goods made to customer
Sale Invoice	Evidence for revenue received
Electronic Money Transfer Reference #	Receipt of money from customer
Credit Note/Debit Note	Adjustments to customer account

Source Documents for Expenditure Cycle

Source Document	**Representing Transaction**
Purchase Requisition	Need for material purchase/service procurement
Purchase Order	Evidence for order placed with vendor
Goods Receipt Note	Evidence for goods/services received
Electronic Money Transfer Reference #	Payment to vendor
Credit Note/Debit Note	Adjustments to vendor account

Source Documents for Human Resources Activity/Payroll Cycle

Source Document	**Representing Transaction**
IT Declaration Form (called Form W4 in USA)	Collection of tax-related inputs to tax recovery from salary
Time Sheets	Evidence of the time worked by employees
Job Cards/Project Cards	Evidence of the time spent by an employee in a specific job or project

Source Documents for General Ledger Transactions

Source Document	**Representing Transaction**
Journal Vouchers	Evidence of recording the accounting transactions directly in the general ledger

Data Processing Cycle

Batch Processing

As the business transaction events occur the data is captured on the fly and stored. The stored data remains as raw data in the system in case of batch processing, till the time the batch-processing event is triggered. In other words, if the collected data is aggregated and periodically processed, it is called batch processing

Online Real Time Processing

Contrary to batch processing, certain systems can be set up for online real time processing. In this case, each and every transaction is processed by the system as an when a transaction occurs. Based on the business rule set-up, the processed data is updated in the different databases in the system.

All Information Relating to a Transaction

Both in the case of batch processing and online real time processing, the systems are generally set up to capture all information such as who has executed the transaction in the system, who has approved and the areas that have been impacted along with the date and time stamp. In certain cases, the system hardware information such as the terminal from which the update has been made and the IP address of the system which updated the data are also captured. This information is also called as meta data, which means data about data.

System Feedback and Alerts

In general, in online real time processing, the system data input an update and an outcome of the recorded transaction in the system as feedback is provided. For example, if a customer sale order is created, once the order creation is completed, the system responds with the order reference number and a message that the 'order has been successfully created'. In certain cases, based on configuration and capability, an alert may be sent on a specified device to the people who are interested in the transaction.

Audit Trial

Often, the recorded transaction will call for modifications due to business reasons or for rectification of errors. Each of the changes made to the record has to be captured in the system and the entire history of the transaction—from the creation of the record till its archiving—all information has to be stored as system information. This information is automatically updated in the system with the system date and time stamp. This is very valuable information for audit and is called 'Audit Trial'.

Different Types of Files

The data stored in the files are classified based on the nature of the information stored in the file. The type of files can be as below:

1. Master File: In this file, one time information about the repetitive usage of information is stored. For example, in a business, customer information, vendor information, material information, general ledger accounts are often used repetitively. This information is stored in customer master, vendor master, material master and GL Accounts. Changes to master data are very rare but possible.
2. Transaction File: All the transaction information is stored under this category of the files. Sales order, purchase order, material issues are some of the examples of transaction-file information.
3. Reference File: The information stored in this file is often referred to by the program for fetching appropriate data, based on the fulfilment of certain conditions. For example, if the tax to be calculated for sale, the tax has to be calculated based on the geography or state in which the sale is being made. All the states with the corresponding sales-tax information can be stored in a reference file.

2.7 Impact of IT on BPM and Risk of Failure of IT

As mentioned earlier, BPM software suites are available to automate BPM as an IT solution.

These are more 'Process Centric', which means integration of people and processes with systems.

Diagram 2.15 Integration of People, Process and Systems

IT-enabled BPM offers the following capabilities to an enterprise:

1. A close working of business and IT in developing an IT-enabled BPM
2. Ability to integrate people and processes with systems and automation
3. Ability to simulate various process options and choose the most optimal one
4. Ability to monitor and control the processes in real time
5. Ability to continually improve, without much of effort

Advantages of having a BPMS

1. **Automation of Business Processes:** The first and foremost advantage is the automation of

repetitive business process. An automated environment helps an enterprise in having system-driven activities rather than manually driven activities.

2. **Integration with Existing Applications:** BPMS can be integrated with existing applications—including legacy applications. This enables the enterprise to not to lose the investment in an existing IT-application environment.
3. **Cost Savings:** Automated BPM is integrated with existing applications, helping an enterprise to achieve cost savings. The primary reason for cost savings is that automated processes help in a speedier completion than the manual process, thus investment in BPMS assures ROI for an enterprise.
4. **Helps in Documentation needed for Quality Certification**: Quality certifications such as ISO, CMMI, etc. call for a maintenance of a good amount of documentation and evidence that the enterprise is adhering to the designed process. Implementing BPMS helps an organisation to automatically provide the documents and evidence for compliance and recertification.
5. **Redeployment of Human Resources:** By implementing BPMS, the employees of the enterprise are freed from an avoidable manual process, which enables them to engage in more intellectual activities.

Business Risks due to Failure of IT

Applying Information Technology in automating BPM is not successful in all enterprises. There are risks involved in implementing the BPM software suites. The primary reasons are

- Inadequate investments
- Inadequate training to personnel
- Lack of management policies on the protection of data integrity

Other reasons for failure include:

1. Lack of management commitment
2. Ineffective project management
3. Inefficient process-gap analysis and design
4. Software limitations
5. Failure of the enterprise in identifying future needs
6. Inadequate assessment about enterprise transformation management
7. Very old legacy systems not compatible with the new BPMS
8. Over-engineering of processes
9. Unrealistic expectations from BPMS
10. Technology changes

Information: An Asset for an Enterprise

Information is a critical asset for an enterprise. Information helps the enterprises to take strategic decisions and therefore enables them to not only compete in the competitive market but also to survive.

In a global enterprise, where the enterprise operations are geographically dispersed, data is generated in each part of the world. All this data is stored in either locally hosted servers or in a centrally located data centre server, In either case, this data is further aggregated and processed in different analytical processing systems such as Customer Relationship Management, Data Warehousing Systems, Business Intelligence Systems, etc. Using analytical processing systems, the enterprise data is converted into trending information and by applying statistical methods and operation research models enterprises derive meaningful strategic information to direct their business operations.

Of late, besides aggregating the data gathered within the enterprise, data generated outside the enterprise, which is meaningful for the enterprise business decisions, is also leveraged. This is known as 'Big Data'.

Section 3

Business Process Reengineering and Approaches

3.1 Background

Implementing Information Technology for business process automation (BPA) accelerates business benefits derived by an enterprise. When an enterprise applies Information Technology in its operations, there are unique advantages that can be leveraged which are not available otherwise when the processes are being manually executed.

By automating the existing manual processes, an enterprise may miss out on the advantages offered by Information Technology. In addition, the limitations existing in the current manual processes will remain.

In order to derive maximum benefit from BPA, enterprises engage in redefining their processes and in reengineering them, before deploying IT and embarking upon BPA. This exercise is called Business Process Reengineering.

3.2 Business Process Reengineering (BPR)

As defined earlier, **Business Process Reengineering is the fundamental rethinking and radical redesign of processes for achieving dramatic improvement in the critical contemporary measures of performance such as cost, quality and speed.**

Critical words in the definition are

Fundamental rethinking is questioning the way we perform the activities of the business by asking 'What is the activity?', 'Why it is performed?, 'Why it is performed, the way it is currently performed?', 'How it can be done differently?, 'How it can be made as a value addition to the customer?', 'How it can be automated?', 'How it can be made more efficiently?' and so on.

Radical redesign: Without attempting to improve the existing process, it is the design of a new process, looking from the perspective of how it helps customers and/or enterprise and/or employees and designing a new process afresh.

Dramatic improvement: Unlike the programs like 5S, TQM, CMMI, ISO, Benchmarking where the improvements are in the range of 5% to 10%, in BPR the improvement is substantially high—as much as 70% to 80% or even more—in terms of defect prevention and cost savings.

In implementing BPR, an enterprise may or may not implement the IT software tools.

Following are the critical success factors for BPR implementation:

1. **Enterprise-Wide Commitment**: This is a very important factor as an effort calling for a radical redesign of the processes will inevitably cut across all functions at all levels of the enterprise. Therefore, viewing the process beyond a functional perspective and understanding the process end-to-end by all members in the enterprise is very important and this can be achieved only through an enterprise-wide commitment.
2. **People and Skill**: People make a big difference to any effort to make it success or break it as a failure. BPR implementation needs a formation of cross-functional teams within the enterprise and also the engaging of external consultants and experts to lay new radically redesigned processes. The people with appropriate skills are the second important factor.
3. **Understanding of Business and Analysis:** The third factor is a deep understanding of the business and core competence of the enterprise and which unique aspect makes the enterprise different from the rest of the players has to be properly understood by all involved in the BPR implementation.
4. **Investment in an IT Environment:** Successful BPR implementation has to experience the prerequisite of enterprise investment in IT infrastructure and the entire environment. An

adequate investment in hardware, network and software is critical for a successful BPR implementation.

5. **Effective Enterprise Transformation:** As one can understand when a fundamentally rethought radical redesign of process has taken place, the enterprise's way of conducting business transactions is dramatically changed. But, it is natural for people to resist change. In order to achieve a successful implementation of BPR, the enterprise has to plan and execute an effective enterprise transformation across all levels of the organisation.
6. **Continuous Improvement:** Finally, setting up a culture of continuous improvement is part of BPR implementation. As the business continuously experiences changes both inside and outside business, in order to respond to them, there will be a need for continuous changes to be made to the designed processes. Hence, a continuous improvement is part of successful BPR implementation.

3.3 Process-Mapping Approaches

Process mapping is a pictorial representation of the series of activities and their inter-relationship in a process. Evolving from engineering, computer engineering has adopted its best practices. As part of the efforts in system design and in analysing the requirements of system features, there are a few approaches taken by system analysts and solution architects. Here are a few approaches listed below, which are very useful for accountants and auditors in understanding the processes and the activities therein.

1. Entity Relationship Diagram (ER Diagram)
2. Data Flow Diagram (DFD)
3. System Flow Diagram (SFD)
   a. System Outline Charts
   b System Flow Chart
   c. System Run Chart
   d. Program Flow Chart
4. Decision Trees
5. Decision Tables

Entity Relationship Diagram (ERD)

In a process there are several 'Entities' involved in performing activities at different stages and in completing the process. People playing different roles and performing activities of a process are called as 'Entities'. Also, an entity may be an object or a concept.

For example, an entity can be a Customer, Vendor or Provider where the customer, vendor or provider could be an individual or an artificial entity.

An entity can also be an automobile, book or a computer, where all these are objects.

Similarly an entity can be a service such as 'Teaching', where the service is a concept.

As the name suggests, ERD is about explaining the relationship between entities.

The relationship between entities can be one of the following:

- One-to-One (1:1): In this case, for every one record, there is a corresponding one record only. It is also called as that for every parent record, there will be only one child record; For example, location and postal code.
- One-to-Many (1: M): In this case, for every parent record, there will be several child records. For example, one postal code for many streets.
- Many-to-One (M:1): In this case, two or more parent records for a child record. For example, multiple residents on a single address.

- Many-to-Many (M: N): Many parent records have many child records. For example, many suppliers may supply many materials/parts. Suppliers A, B, C may supply material or part numbers M, N.

When an ERD is developed, separate symbols are used to represent an entity, attribute of the entity, key attribute of the entity and relationship. These are provided as below:

Element in ER Diagram	Represented by
Entity	*Entity Name*
Attribute	*Attribute*
Key Attribute	*Key Attribute*
Relationship	*Relationship*

Here is an example of ERD:

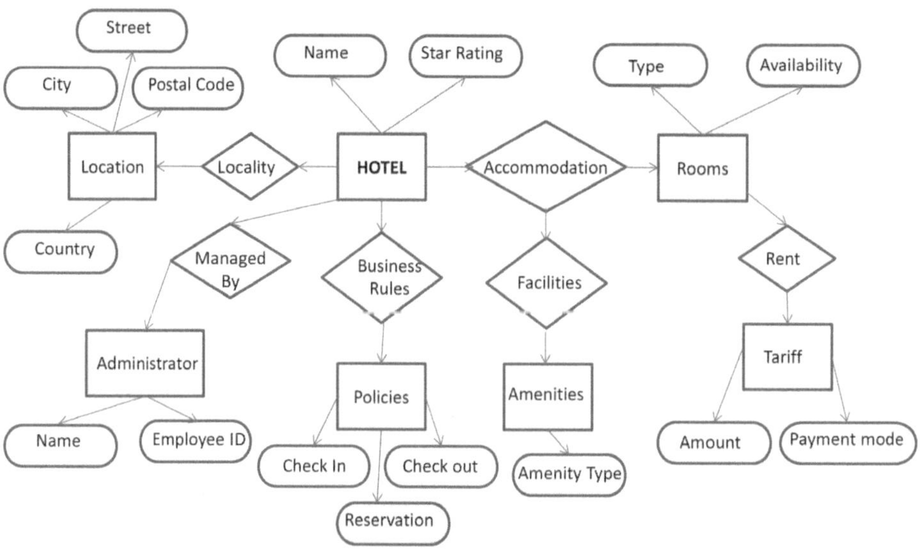

Diagram 3.1 Entity Relationship Diagram

In this diagram Hotel, Location, Rooms, Cost, Policy, Facilities, Manager are the entities. Each of these entities has attributes and some of them are Key-Attributes.

The relationship between these entities is explained with the relationship symbols.

Data Flow Diagram (DFD)

Data Flow Diagram visually represents how a data flows throughout the system.

DFD also explains which entities trigger the process and thus provide data; how the data is processed and where the data is stored.

The elements in a DFD are

- Entity: Entities involved in the process are depicted in a DFD.
- Processes: The processes through which the data flows are explained.
- Data storage: Where the processed data is stored is also depicted in the DFD.

Symbols of DFD

Symbols	Elements
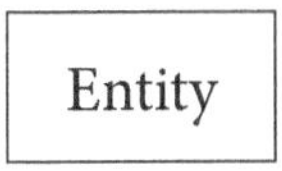	Entity
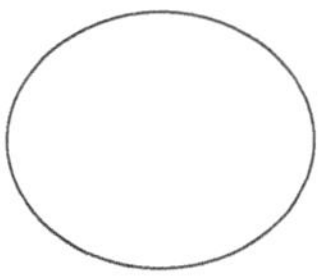 Or 	Processes
 Or 	Data Storage
	Data Flow

Note: There are more symbols in use when developing the DFD; the above are a few of the several symbols and not the exhaustive list.

Example of a Data Flow Diagram

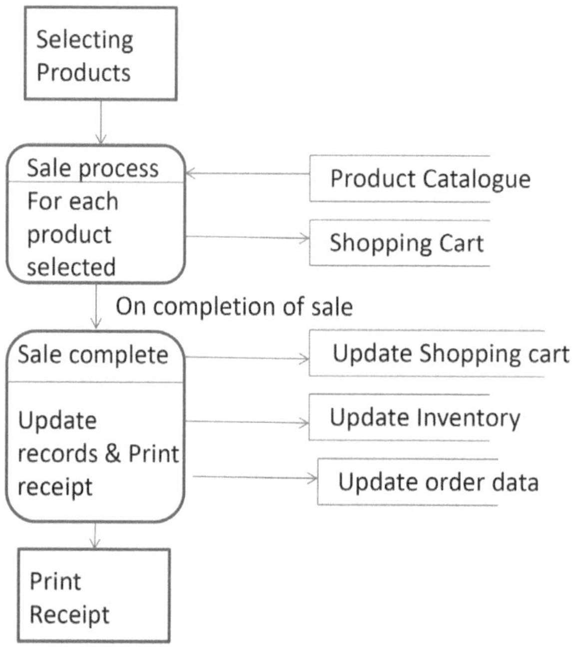

Diagram 3.2 Data Flow Diagram

'Selling items' ('products') and 'printed receipts' are entities.

'Processing each item sold' and 'printing receipts' are the processes.

Data is stored in Updated Basket, Item Files, Updated Item Files, marking the items in the basket as paid and issued to ensure that the basket is empty.

System Flow Diagram

System Flow diagram is also called as flow charts. These are diagrammatic representations of the steps in a process from the beginning till the end.

As part of process development, flow charts are drawn and explained. The flow charts are also used as reference documents while executing a process. Software engineering leverages process flow charts for determining the system requirements. Flow charts help in (a) simplifying the complex fuzzy logic (b) including all steps needed for a process to complete the work.

Symbols of Flowcharts

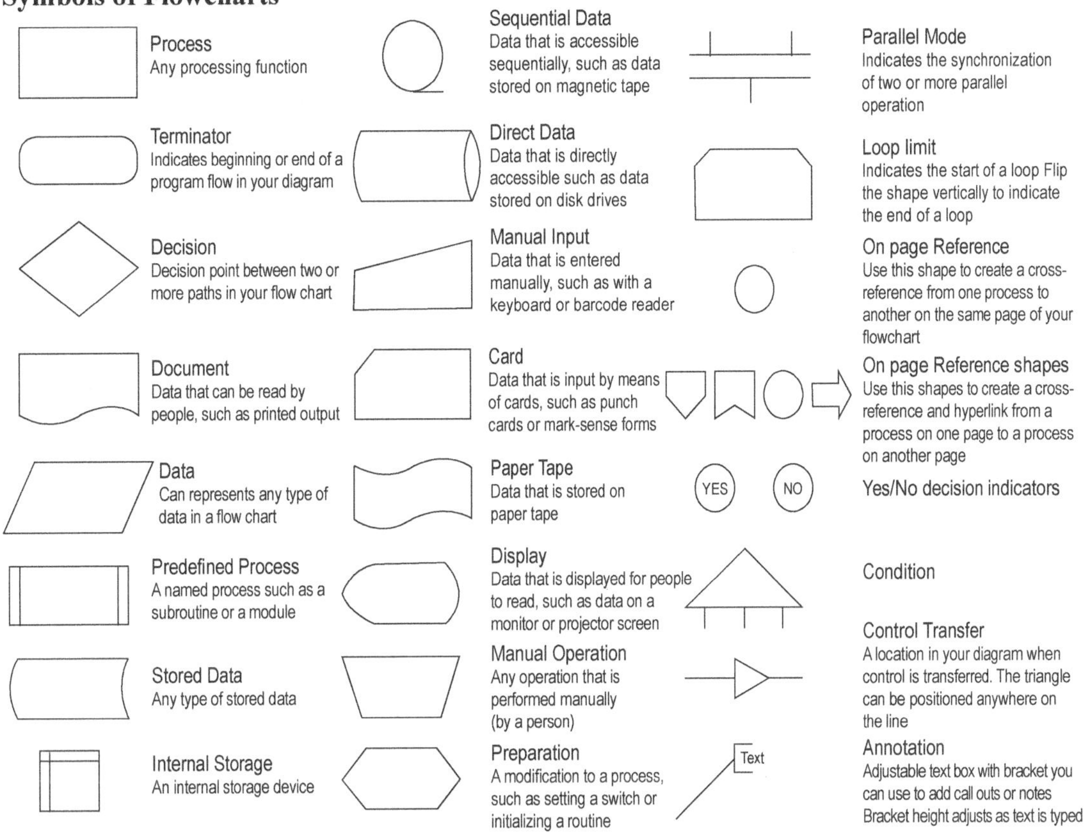

In software engineering the following flow charts are generally used:

1. System Outline Chart
2. System Flow Chart
3. System Run Chart
4. Program Flow Chart

System Outline Charts

System outline charts are used to provide a summary of the system functionality outlining inputs, process and outputs of the system. This document is critical in software documentation as there will be several systems that will be running in an enterprise and a summary of each of the systems is important to have for explaining the functions of the system and interaction between the systems. System outline charts are a subset of enterprise architecture.

Here is the example of a system outline chart:

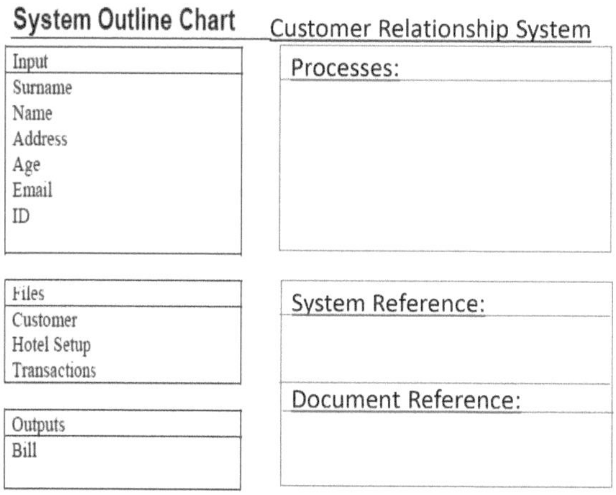

Diagram 3.4 System Outline Chart

System Flow Chart

A system flow chart explains how a system is working and what are the activities performed by a system. The activities are explained step by step. Here is an example of system flow chart for an online departmental store on inventory management:

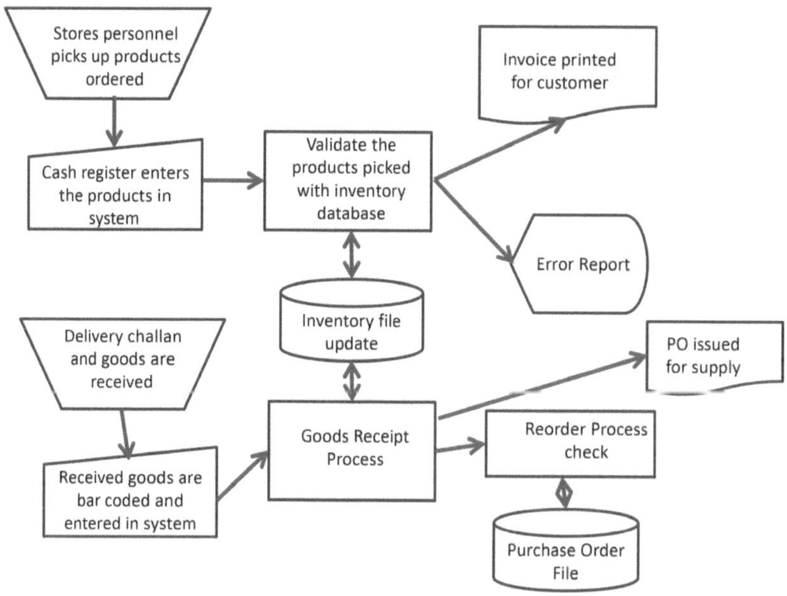

Diagram 3.5 System Flow Chart

System Run Chart

System run chart is a subset of the System Flow Chart. It shows how the computer operations are performed when each time the program is run (or executed). System Run Chart includes inputs, outputs and files. Example of a run flow chart for an INR currency value conversion to US$ processing system is as follows:

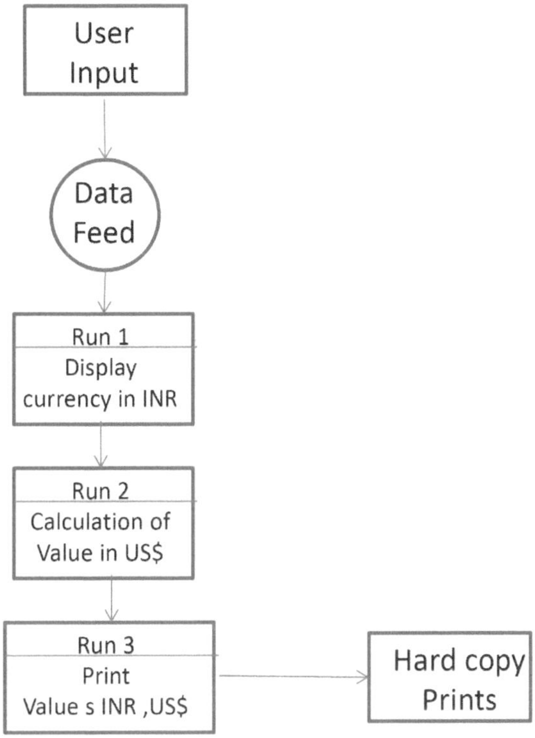

Diagram 3.6 System Run chart

Program Flow Chart

Program flow chart explains in detail all the steps performed within a system to convert the input into output. The various steps performed by the system are logical, arithmetical and algorithmic. A program flow chart is very useful for translating the program requirement into a diagrammatic representation commonly understood by both system analysts and programmers.

Here is an example of a program flow chart, where the program displays the highest number of the given three numerical values:

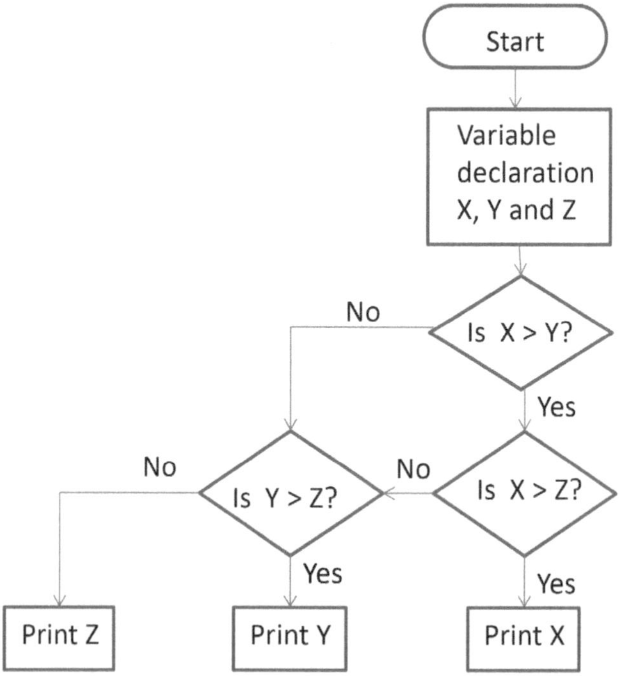

Diagram 3.7 Program Flow Chart

Differences between DFD and System Flow Diagram

Data Flow Diagram	**System Flow Diagram (Flow Charts)**
DFD represents flow of data	Flow chart explains step-by-step activities for completing a process
DFD shows the path of data from external source to internal system and vice versa	Flow charts do not have any input from external sources nor is there any output to external sources
The way data is processed is not explained in DFD; it merely explains flow.	The sequence of activities and process is explained in the Flow Chart
DFD explains the functionality of a system	Flow Chart explains how to make a system function or explains the features of a system
DFD explains the flow of data that is processed during the completion of a process	Flow chart explains how to design a process
DFD can be of the following types: 1. Logical data flow diagram 2. Physical data flow diagram	Flow Charts can be of the following types: 1. System outline charts 2. System flow chart 3. System run chart 4. Program flow chart

Decision Tree

The analysis of a complex decision with a significant uncertainty can be confusing. The reasons for confusion are

1. Probability of uncertainty and the outcome of an uncertain factor
2. Simultaneous impact of multiple uncertain factors impacting the outcome

Decision tree is an approach to constructing a diagrammatic presentation of the possibility of occurrence of the decision-making factors and the corresponding outcome. It is an alternative to the statistical method approach to decision making. Here is an example of conducting surgery for a patient by a doctor, provided all indicators under biological, radiological and pathological tests are either acceptable or positive. The possibility of various factors being positive, acceptable and negative can be tabulated in a tabular form and the decision outcome can be predicted.

Indicators	Favourable indicators	Acceptable	Adverse indicators	Decision
Biological test	Yes			Yes
Biological test		Yes		Yes
Biological test			Yes	No
Radiological test	Yes			Yes
Radiological test		Yes		Yes
Radiological test			Yes	No
Pathological test	Yes			Yes
Pathological test		Yes		Yes
Pathological test			Yes	No

Predictors Target

If the same data tabulated above is represented in a decision tree, it will be as follows:

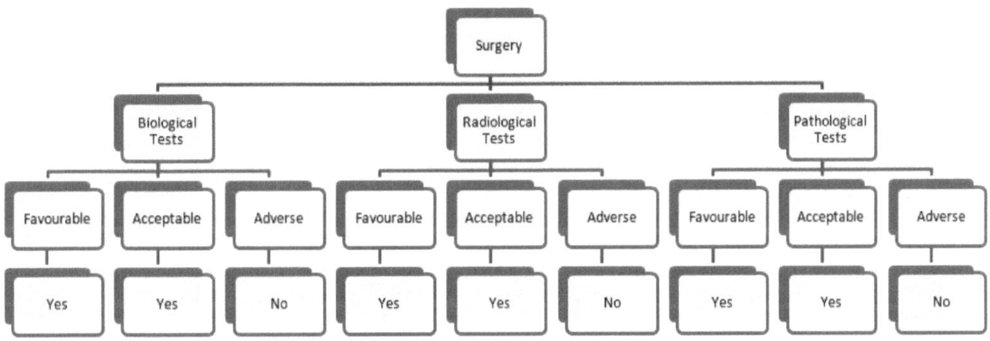

Diagram 3.8 Decision Tree

In the decision tree, logical operators "AND", "OR" are used to replicate the structure.

In this example, the logical operator is AND in the sense that the surgery can be conducted only if all the tests are either with favourable indicator or acceptable indicator.

In the decision tree, one can find branches and decision nodes. In the above example, biological test, radiological tests and pathological tests are branches and the last level decision nodes (Yes or No) are called leaf nodes. Uppermost nodes are called root nodes and in this case "Surgery" is the root node.

Decision Table

Decision tables are a precise yet compact way to model the complex logic. Decision tables are like flow charts; they help in designing the process and a system. Decision tables are leveraged by software engineering for system design and program design.

Decision tree provides a schematic view of the inference process of a decision-making process. Each decision rule of a DT is composed of a premise (condition) and a conclusion (action).

A decision table helps in decomposing the complex logic into a compact table which shows that 'for a given conditions' fulfillment, what should be the decision'.

Structure of a Decision Table: There are four quadrants in a decision table as depicted below:

Conditions	Conditions Alternatives
Actions	Actions Entries

Diagram 3.9 Tabs in a Decision Table

Like flow charts, decision tables are also helpful in offering clarity to a programming logic. Consider the examples given below:

Example with Simple Conditions:

Temperature data of three different days of various locations have been provided and the system is expected to find out the day that had the highest temperature for each location.

Condition	**Rule 1**	**Rule 2**	**Rule 3**	**Rule 4**	**Rule 5**
Day 1 and 2 compared, 1is higher-C1	Yes	Yes	No	No	No
Day 1 and 3 compared, 1 is higher-C2	Yes	No	No	No	No
Day 2 and 3 compared,3 is higher-C3	-	-	-	Yes	No
Action					
1 day higher-A1	X				
2 day higher-A2			X		X
3 day higher-A3		X		X	

In the above table

Condition Statements: Statements suffixed with (C1), (C2) and (C3) are condition statements

Action Statements: Statements suffixed with (A1) through (A3) are action statements

Condition Entries: Entries in columns 1 through 5 against condition statements are called condition entries

Action Entries: Entries in column 1 through 5 against action statements are called action entries

Condition Identifiers: (C1), (C2) are called condition identifiers

Action identifiers: (A1) through (A3) are called action identifiers

Rules: A unique combination of conditions and actions that corresponds to conditions is called rules

Rules Identifier: Row header 'Rule 1' through 'Rule 5' is rule identifier

Interpretation of the Table

Step 1: Day 1 temperature is compared with Day 2

Step 2: Day 1 temperature is compared with Day 3

Step 3: Day 2 temperature is compared with Day 3

Possible results:

Day 1 is higher than Day 2 and Day 3—Day 1 is highest

Day 1 is higher than Day 2, but less than Day 3—Day 3 is highest

Day 1 is less than Day 2, but higher than Day 3—Day 2 is the highest

Day 1 is less than Day 2 and Day 3. Then Day 2 is compared with Day 3 and Day 3 s higher—Day 3 is highest

Day 1 is less than Day 2 and Day 3. Then Day 2 is compared with Day 3 and Day 2 is higher—Day 2 is the highest.

Verification of the Completeness of a Simple Decision Table:

The following steps are followed to verify whether a decision table is complete and correct.

The meaning of 'complete and correct' means that every data that is filtered through the decision table conditions will result in at least one action.

First: Count the number of conditions—In the above example we have 3 conditions—C1,C2 and C3

Each conditions have either 'Yes' or 'No' answers—two possible results.

Compute the value of 2 to the power of (number of conditions) = $2^{\text{number of conditions}} = 2^n$

In the example we have three conditions and therefore, $2^3 = 8$

Second: Consider each rule in the conditions section. In each rule, how many conditions have been skipped is to be determined. The skipped conditions are marked with hyphen **'-'**

For each skipped rule 2 results are possible—"yes" or "no".

For each rule compute value by making the number of skipped conditions as an exponential of 2. If one condition is skipped then 2^1 and if 2 conditions are skipped then 2^3; For 3 skipped steps it will be 2^3 and so on.

In the above example,

For Rule 1 = there is one step skipped or with '-' and therefore $2^1 = 2$

For Rule 2 = there is one step skipped or with '-' and therefore $2^1 = 2$

For Rule 3 = there is one step skipped or with '-' and therefore $2^1 = 2$

For Rule 4 = there is no step skipped or with '-' and therefore $2^0 = 1$

For Rule 5 = there is no step skipped or with '-' and therefore $2^0 = 1$

Sum all the results = 2+2+2+1+1 = 8

If $2^{(\text{number of conditions})}$ = Sum of $2^{(\text{number of skipped conditions})}$ for all the rules, then the table is considered to be complete and correct. If not, then the table needs improvement.

For a complete & correct decision table, $2^{(\text{number of conditions})} = \sum_{i=1}^{n} (2^{(\text{number of skipped conditions})})$ where n = number of 'Rules'.

In the above example, $2^3 = (2^1+2^1+2^1+2^0+2^0)$, therefore the decision table is correct and complete.

Note: The above verification is possible only for decision tables with simple conditions and there are advanced methods for complex multiple conditions, with AND, OR, ELSE operators.

Example with Complex Multiple Conditions:

In an enterprise there are two types of employees: Full-time hire and hourly rate.

Full-time hires will be eligible for annual vacation, but not the hourly rate employees.

Therefore, for full-time hires the payroll to be processed with the base salary provided the vacation does not exceed the maximum number of the annual vacation-paid time-off of 10 days.

For all hourly basis employees, payroll is to be processed for the number of hours worked. Against the weekly hours of 40, if any of the employees have worked more than 40 hours, then overtime salary has to be paid and overtime report has to be produced.

If any of the full-time hire has worked for less than 40 hours, then an absence report has to be produced.

If this scenario has to be explained to a programmer or developer, the logic may appear to be very complex. If the above complex logic is decomposed in a decision table, it will be easier for the programmer to develop the code. Here is the decision table for the above scenario:

Condition	Rule 1	Rule 2	Rule 3	Rule 4	Rule 5	Rule 6	Rule 7
Full time employee? (or Contractor)-C1	Yes	Yes	Yes	Yes	No	No	No
Worked 40 hours?-C2	Yes	No	No	No	Yes	No	No
Worked >40 hours?-C4	-	-	-	Yes	-	-	Yes
Worked <40 hours?-C3	-	Yes	Yes	No	-	Yes	No
This year paid time off is exceeding 80 hours?-C5	-	Yes	No	-	-	-	-
Action							
Process salary for 40 hours a week-A1					X		X
Process base salary-A2	X		X	X			
Process proportionate salary-A3		X				X	
Process overtime salary for hours more than 40 hours-A4				X			X
Print Absence report-A5		X	X			X	
Print Overtime report-A6				X			X

Interpretation of the Table

Step 1: Type of employee for whom payroll to be processed has to be determined by the system—Full-time hire or hourly employee (contractor).

Step 2: No. of hours worked by each employee has to be determined—whether 40 hours or more than 40 hours or less than 40 hours.

Step 3: If no. of hours worked is 40 hours or more then process the base salary for full-time employees and salary for 40 hours per week for hourly employee (contractor).

Step 4: If no. of hours worked is more than 40 hours then process over-time for both full-time employees and hourly employees (contractor) as per rule

Step 5: If no. of hours worked is less than 40 hours, perform step 5A or 5B as the case may be

Step 5A: If employee type is hourly employee (contractor) then calculate salary for the number of hours worked.

Step 5B: If the employee type is full-time hire, then the no. of hours of paid time-off availed this year has to be determined and process 5BA or 5BB, based on the no. of paid time-off availed during this year.

Step 5BA: If the employees have worked for 40 hours and less and if it is 40 hours and paid time-off availed is less than 80 hours in the year—process base salary

Step 5BB: If worked less, then 40 hours and paid time-off is more than 80 days in the year—process payroll with the proportionate base salary for the hours worked.

Section 4
Computer Systems and the Auditing IT Environment

4.1 Background

Next to the Industrial Revolution in the early part of the last century, the Information Technology revolution has been experienced by the world. The Information Technology revolution has been further accelerated by the Internet revolution.

In today's business world, Information Technology has been the basic ingredient for doing business. In fact, the application of Information Technology and information systems has transformed today's business, making business operations more effective and efficient.

Besides the application of Information Technology in business, new business models have been developed with IT as the key component of business. For example, all IT-enabled services such as IT services, Business Process Outsourcing are modelled around Information Technology.

Internet technologies have further strengthened the ability of IT applications in business. In addition there are businesses, which have been modelled around internet technology: for example, Google, Yahoo or any e-Commerce business.

In the early evolution of Information Technology, enterprises were automating functional processes. Over the years, through radical thinking, new business processes have been designed, leveraging IT through concepts such as Business Process Reengineering and the enterprises have immensely benefited with substantial cost savings along with speed in processing.

While technology such as Networking helped enterprises to interface different applications within an enterprise; Internet technology has helped enterprises to interface their respective IT environments through a methodology called Business-to-Business (B2B).

As more and more enterprises embrace the latest technologies, developing new business models, their impact on accounting and audit professionals is becomes unavoidable.

There is a strong need for both the accounting and audit professionals to develop a deeper understanding of the current and emerging technologies and their applications in business along with how this impacts the accounting system—in turn impacting on the valuation of assets and business.

Only through such a deep understanding of technologies, can an accounting and audit professional make a judgement of how the enterprise financial statements are prepared through the culmination of an automated IT environment through an inter-enterprise data flow, and that such financial statements represent a true and fair view.

This section explains the following fundamentals of IT:

1. Computer System
2. Hardware

3. System Software
4. Application Software
5. Databases

In addition, it covers the auditing of a business in an IT environment.

4.2 Need for Information Technology

Irrespective of the size of the enterprise, every business deploys IT in conducting business operations. The main reason for the application of IT in the past was to ensure speed and accuracy in their business operations. Over the years, the Information Technology application has been the strategic differentiator for enterprises. While the early adaptors of IT applications reaped huge benefit, overtime, application of IT has become a basic need for almost every business.

In the past, enterprises were viewing IT applications as overhead. In the current age. IT is considered as a strategy accelerator and enabler.

Matured enterprises maximise their IT operations to optimise the enterprise resource management. Thus, these matured enterprises are able to quickly respond to the rapidly changing environment by evolving suitable strategic responses.

Information Technology helps enterprises in

- Data and information management
- Communication management
- Automated Business Process Management

Data and Information Management

The usage of Information Technology for data and information management is not new. For ages, computers have been applied for data and information management.

However, the technological advancement in applying computers and storage devices for data and information management is new and there is a dramatic reduction in cost of such operations in storing data and information.

Independent of technological advancement, newer applications of computers in business are emerging, which handle data and provide information. For example, since the age of functional automation, computers have been applied in handling transactions in a business. As part of the business process evolution, Enterprise Resource Planning (ERP) software packages have been developed. Most of the ERP packages are popularly known as Online Transaction Processing system (OLTP). Over the past couple of decades enterprises have amassed transaction data and they have found value for these massive data and thus came the new applications called Online Analytical Processing system (OLAP). OLAP offers a diagnosis of business trends.

As digitisation advanced, in the current era, documents came to be stored in the digital format and managed through the document management system, which offers solutions for managing the document management life cycle.

As data and information, including documents, are stored, in digitised form, there is a strong need for security of the IT environment for every enterprise. Thus, the security solutions such as access control, role-based authorisation, single sign-on, cryptography, etc. have come of age.

IT and Communication Capability

As technology advanced, its advancement resulted in the convergence of Information Technology, communication and entertainment in single media and device. This was possible through Internet technologies. Communication capability was developed using the Internet technology through a protocol called Voice over Internet Protocol (VoIP).

Example of VoIP-based telephony services are Skype, Google Voice and Vonage telephony services.

Besides the voice communication over Internet, the Internet is applied for collaboration through the virtual model. Here are some examples of virtual models:

Microsoft SharePoint Portal—for virtual collaboration

IBM Lotus Suite of products—for instant messaging, mail, collaboration

WebEx—for virtual meeting

Saba—for virtual classrooms and training sessions

4.3 Computing

There are various definitions of computing as defined by different experts in the field. The Association for Computing Machinery (AMC) has defined computing as *"a goal-oriented activity benefiting from or creating computer"*.

The definition broadly covers two goal-oriented activities: (1) Creating a computer (2) Any activity benefiting from computer.

Hence, computing includes the following:

1. Designing a computer
2. Enhancing the capability of computers
3. Designing software for computers
4. Applying computer for an activity
5. Using computers for data processing
6. Using computers for communication
7. Applying computers for entertainment
8. Applying computers for educational needs

The list is as endless as the present world; computer application is present in virtually in all aspects of life.

The sub-disciplines of computing are (1) Computer Science (2) Computer Engineering (3) Information System (4) Information Technology (5) Software Engineering.

Computer Science

Computer Science is about a scientific and practical approach to computation and its application.

It involves a systematic study of feasibility, structure, expression and mechanisation of methodical processes (or algorithm).

Computer science is a methodical process that underlies the acquisition, representation, processing, storage, communication of and access to data/information.

This information may be encoded in bits and bytes in a computer memory or transcribed engines and protein structure of a human cell.

In a nutshell, Computer Science is computation and design of the computational system.

Computer Engineering

Computer engineering refers to the discipline of integrating electrical engineering and computer science.

Computer engineering is required to develop computer hardware and software.

Computer engineering involves the design of individual micro-processor or Information Technology include technologies personal computer or super computer, which includes circuit design.

Importantly, computer engineering involves the integration of all subsystems for the creation of a holistic system.

Information System

Information system can be defined as the study of a complimentary network of hardware and software that enterprises, employees and individuals use to collect, process, store and distribute data/information.

An information system supports enterprise operations, management and decision-making.

Information System integrates the enterprise processes with people and technologies—Information and Communication Technologies.

Information Technology

Information Technology is the study, design, development, application, implementation, support or management of computer-based information system.

Information technology includes technologies related to networks, software development, installation, deployment, planning and managing technology cycle of enterprise and maintenance of hardware and software (upgraded or replaced).

A broader definition of IT includes communication technology.

Advancement of IT resulted in a convergence of information, communication and entertainment technologies; IT also includes entertainment technology, such as gaming software.

Software Engineering

Software engineering is defined as a systematic, disciplined, quantifiable approach to design, develop, operate and maintain software and the study of these approaches.

Software engineering also includes the application of engineering principles to software development.

Software engineering knowledge is critical for auditors to familiarise themselves with IT and IS and their practical deployment.

4.4 Technology of Computing

The technology of computing is a vast subject. Some of the important concepts are covered here.

Server

- A server is a computer. As this computer performs the activity of serving the clients (client is another computer used by end-users), it is called server.
- From a hardware perspective, a server is a hardware device on the network dedicated to run one or more services. The server is also called "Host", as a server hosts different services.
- From a client-server architecture perspective, a server performs computational tasks on the behalf of clients. Servers are often dedicated. In certain scenarios, a single computer may perform the activities of both client and server. In other words, client and server may be in a single device.
- Essentially, it is not the size of the computer, but the tasks performed that determine whether a computer is a server or not. Generally, servers are set up with high performance computers with a high-speed computing capability and data storage capacity as they needs to serve several clients simultaneously.

Types of Servers

Though a server is capable of performing different services, in practice, a server is dedicated to performing a particular service. This is decided based on the volume of the workload and capacity of the server. When a server is dedicated to the performance of a particular service, it becomes a type of a server. Here are various types of servers:

File server: As the name suggests, the file server is for storing files. Any user in the network (subject to access and authorisation) can store files in this server.

Network server: A network server manages the network traffic. In a client server environment, several clients (users) issue commands to the server and the volume of commands is called 'traffic' in the network. A network server typically manages the traffic of the network and handles the traffic based on the rules that have been set.

Mail server: In an enterprise, handling e-mails of the users requires a server. Mail server stores and move mails in an enterprise.

Web server: There will be several strangers who visit the websites of an enterprise. Therefore, there is a strong need for a dedicated server for maintaining websites and webpages. This is required not only from a volume of service perspective but also from a security perspective. A web server serves webpages.

Application server: In a client server architecture, the applications are hosted in a server and the users can invoke the application from the client systems. This server stores applications and serves the clients.

Database servers: In a client server architecture, under the concept of an 'N tier' architecture, databases are set up as a separate tier for storing data. A database server holds databases and stores data.

Computing Architecture

As we are already aware, a computer has hardware; CPU and software are run on a computer to deliver the desired results.

Hardware is a fixed platform capable of executing collections of instructions. Those collections of instructions are embedded in the hardware systems; these are also known as 'Firmware'.

Software is a set of instructions which can be combined as a building block in various permutations and combinations to develop sophisticated programs. These programs are not embedded in the hardware platform—instead, they are stored elsewhere and manipulated in the computer memory—and are similar to that of the data being manipulated in the memory of the computer.

Through different softwares the same hardware platform can be made to behave differently through different sets of programs. For example, when you load the Word application of MS Office the computer behaves as a word processer and the behaviour of the computer is completely different when you load an Excel spreadsheet. Similarly, a computer can be used for airline reservation operations and the same computer can be made to process a laboratory information management system in an enterprise in the chemical industry.

Central Processing Unit (CPU) is in charge of executing the instructions based on the program that has been loaded; software programs tell the CPU what arithmetical operations are to be performed or what logical activities are to be executed and when to move on to invoke another program and where to write the results. The CPU is the centerpiece of the computer system. It interfaces between hardware, software and other elements of the computer architecture.

Computer architecture is a conceptual design and fundamental operational structure of a computer system. It also includes technical drawings and a documented functional description of the architecture.

There are three major components of computer architecture:

1. Instruction set architecture (ISA)
2. Micro architecture
3. System design

Instruction Set Architecture (ISA)

The characteristics of ISA were developed by IBM in 1964, when it developed the IBM 360 system. ISA is an abstract model of computer system written in assembly-level programming language. ISA includes instruction sets, memory address mode, processor registers and addresses and data formats. ISA is about defining and designing what the chip has to do. It is a bridge between the computer hardware and software. ISA tells the computer what to do and how to do.

Classification of Instruction Sets

Complex Instruction Set Computer (CISC): CISC is a control unit. In this instruction set there are a number of microelectronic circuitries. CISC generates a set of control signal circuitry activated by a micro code. An example of CISC is Intel processors.

Reduced Instruction Set Computer (RISC): RISC is also a control unit. Unlike CISC, RISC has fewer and separate electronic circuitries. It is a hardwired control unit and takes less power as compared to the CISC. Since RISC is having fewer circuitries and a small instruction set, codes have to be written into the software program for complex computation. A good example of RISC is IBM RS6000.

Micro Architecture

Micro architecture is a still lower level of architecture of a computer system with all detailed descriptions about the operations of all parts of the computing system. Micro architecture also includes the details about inter-connections and inter-operations of the different parts of the computer system for implementing the ISA. In other words, it is about what an ISA has to do and how it is to be done.

System Design

Buses and Switches interconnect components inside the computer and transfer data between them components. Certain external buses and switches interconnect two or more computers and software.

Memory Controller Hierarchy is a digital circuit unit that manages the flow of data going in and coming out from the main memory. The memory controller can be a separate chip or integrated with the other chips.

Direct Memory Access (DMA) is a new feature in modern computers, where the hardware subsystem within the computer system can access system memory independent of the CPU. This is accomplished through having additional processors along with the main processors. In this case, the main processor is called host processor and additional processors are known as co-processors.

Multi-Processing, Virtualisation are relatively newer concepts which extend the system capabilities. Multi-processing is the use of two or more processing units (CPU) within the single computer. Virtualisation can be hardware virtualisation or platform virtualisation. It is something like having an additional computer with an operating system. Using a real machine, one can access the virtual machine and operate the software.

How does Processing Happen within the Computer?

1. As soon as the system is switched on, the electricity passes through the logic circuitries and the firmware—popularly known as BIOS (Basic Input/Output System)—embedded in the hardware is loaded into the Read Only Memory (ROM) and the computer is ready to perform the computing operations.
2. Then, as one executes a program, the program (a set of instructions) from the hard disk is sent to the Random Access Memory (RAM).
3. Bus that interconnects different components of the system and the CPU, passes the content of the RAM to cache.
4. From cache the instructions are moved to the registers present in the CPU.
5. Control Unit (CU) handles the data to be moved to the Arithmetical Logical Unit (ALU) for processing data and executing instructions and in turn, the Memory Controller handles the output of the processed data provided by the ALU and sends it back to RAM.
6. As the program is executed, fetching data from the databases in the hard disk and writing back results in another database or displaying on the screen or sending to printer for printing are part of the execution process.

The above processes are happening at an incredible speed. Imagine millions of processes being processed in a minute!

The above steps are explained in the flow chart below:

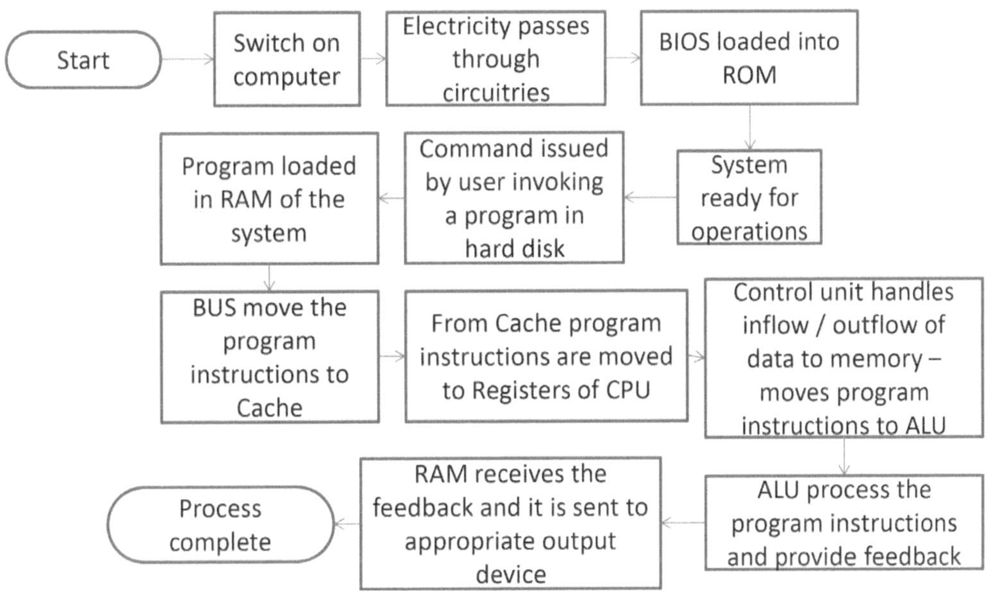

Diagram 4.1 How does a Computer Work?

The process of fetching instructions, decoding, executing and sending feedback (store results) is the cyclical activity performed by a computing system. The diagram below explains how this is done and which component of a computing unit performs each of these activities.

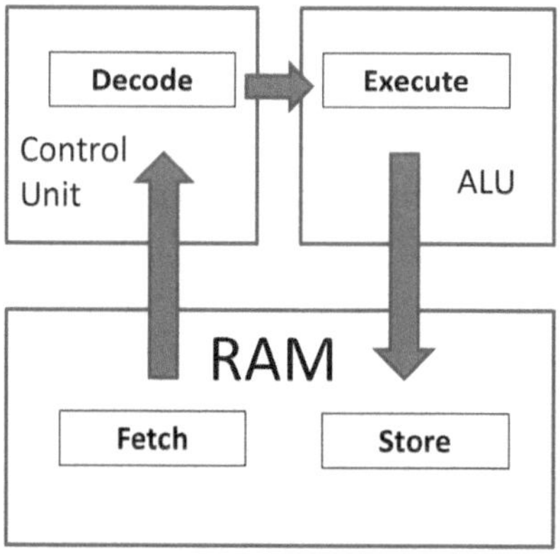

Diagram 4.2 Fetch-Decode-Execute-Store Explained

4.5 Hardware

Hardware of the computers is the one component that we physically we see, touch and experience. Computer hardware can be classified into the following categories: Input devices, processors, output devices and storage devices.

Input Devices: These are the devices through which a user provides commands or provides input data to the computer. Examples of input devices are keyboard, mouse, barcode scanner, MICR readers, webcam, stylus pen, microphone and touchscreen. Through these input devices, we are able provide data input, voice input and image inputs to the computer.

Processing Devices: Inside the computer, a chip contains the central processing unit and memory. The central processing unit—also known as CPU or microprocessors—is the device that interprets the command in the program and executes the commands.

The chip is made of a small silicon wafer, which is equal to the power of several millions of transistors. The transistor switches could be on or off, which corresponds to 0 or 1. The CPU is further classified into three functional units:

1. **Control Unit:** controls the flow of data and instructions to and from memory, interprets the instructions and decides which task is to be executed and when.
2. **Arithmetic and Logical Unit:** This is the unit that performs the mathematical and logical operations in the computer. This is also known as ALU.
3. **Registers:** These are high-speed memory units within the CPU storing data. They could be either 32 bits or 64 bits. Register types are:
    a. **Accumulators** add the running totals of arithmetical values.
    b. **Address Registers** are where memory addresses are stored and tell the CPU where in the memory the instructions are located.
    c. **Storage Registers** are where the data is temporarily stored while processing in the system memory.
    d. **Other registers** are used for general functions.
4. **Data storage devices:** These are the devices where the data and instructions are stored while data processing.
    a. **Internal memory**: Registers and internal memory within the CPU are very small but are of high speed.
    b. **Primary memory:** These are the devices in which any location can be accessed in any order.

The two types of primary memory are

i. **Random Access Memory (RAM):** This memory is alive only when the power is on and the data in this memory is often modified by the system.

ii. **Read only memory (ROM):** This memory is static; often used when the system is booted and the manufacturer of the computer stores boot instructions in this memory. As the name suggests, this memory can only be read and cannot be overwritten. The content of this memory is not lost even when the power is switched off.

iii. **Cache memory**: This memory is to augment the additional memory space to store data between the register and main memory. This memory stores the copy of data from the more frequently used main memory, which enables the data flow aligned with the data handling speed of registers.

iv. **Virtual memory**: This memory is not real memory and part of the CPU. Certain operating systems enable this memory in conjunction with a hard disk. When the RAM needs more space than what is available, the OS allocates some space from the hard disk as memory and this is called virtual memory. The program will be run using the memories of both the RAM and the virtual memory.

v. **Secondary memory:** This is not part of the CPU and is located outside it. When a program is running, the CPU accesses the data in the primary or main memory which is volatile based on the power supply and very small in capacity. But a device is needed to store the data and program, when the power is off and the secondary memory addresses this need. The CPU cannot access the secondary memory directly and uses the input–output channel to do so. Secondary storages are very large in size and also inexpensive compared to the cost of registers. Flash drive, USB drives are examples of secondary memory.

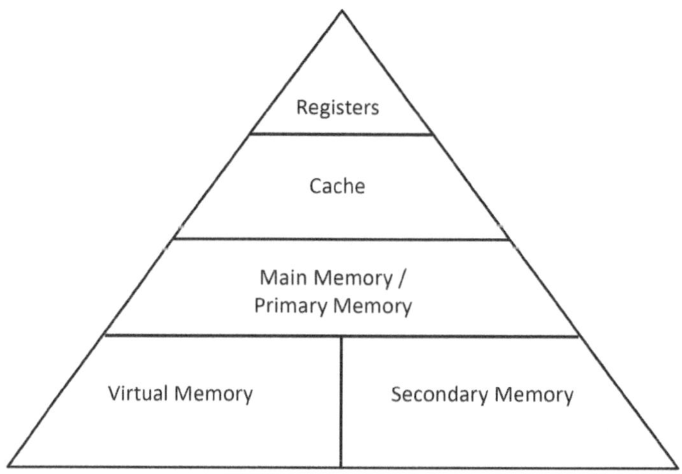

Diagram 4.3 Memory Hierarchy

5. **Output devices:** Once the system processes the data, the result and feedback are sent to the output device. From a business value perspective, a computer performs complex data processing involving statistical methods and operation research and provides valuable inputs to the business decision makers. The output generated by the system, after the completion of data processing, is sent to the output devices. System monitor, speakers, printers are examples of output devices.
   a. System Monitor
      i. **CRT monitors** are those on a technology called the cathode ray tube that uses cathode to emit electrons on the tube to that illuminates the phosphorous that sweeps across the screen rapidly.
      ii. **LCD Monitors** are with Liquid Crystal Display technology.
   b. Printers
      i. **Laser printers** are the ones that uses the technology of a electrostatic process
      ii. **Ink Jet Printers** are the ones forcing ink through nozzles to print images
      iii. **Plotters** are those devices that print high-quality drawings and images with the help of computer-directed pens.
   c. Speakers
      i. **Built-in speakers** are part of laptops
      ii. **External speakers** are generally installed for desktop systems

6. Hands-free headphones are used for portable devices.

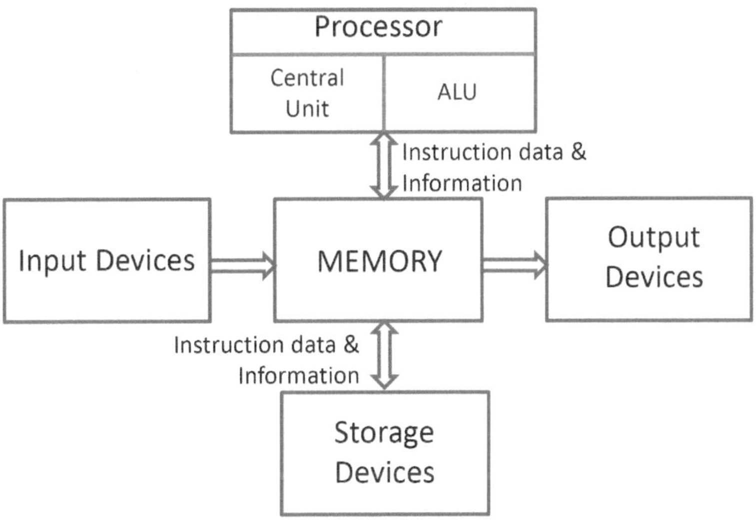

Diagram 4.4 Computer Hardware Architecture

Network Links

Communication is critical for an enterprise, be it voice communication or data communication. In the current era of convergence of communication, information and entertainment, a single network is capable of carrying voice as well as data. Also, for the purpose of data and information-sharing, computers in every enterprise have been inter-connected through computer networks, be it LAN or WAN.

A Computer Network is a collection of computers and other devices/resources interconnected for data and resource-sharing. In a networked environment, one device, through a process executed in that device, can send or receive data from another device. The network can be two categories:

Connection-Oriented network

In case of a connection-oriented network, first the connection is established and then the data is transmitted. It is comparable to telephonic network.

Connectionless network

In case of connectionless network, the data includes the details of the destination, hence, the data long with the address gets transmitted without any connection. It is something similar to postal network.

Basic Aspects of Network

The following are the basic aspects of a network data transfer:

Routing: It refers to how the data is transferred from source to destination.

Bandwidth: Bandwidth means the amount of data that can be transferred over the network in a given time.

Resilience: Resilience is the ability of the network to come back and resume after a temporary disconnection or loss of data.

Contention: Contention refers to a conflicting situation, when two or more systems are trying to connect with each other at the same time.

Benefits of Network

Distributed Information: It is the network-enabled distribution of data and information which accelerates business process reengineering. A data created in any part of the enterprise is made available to the entire enterprise through the network. For example, if the marketing department acquires new customers and a customer master data is created, the same data is available to the shipping department for shipping activities, the accounts department for receivable management, etc.

Resource-Sharing: In a network, besides computers, resources such as telephone, printers, etc. can be connected and all the people connected to the network can share the same.

Computational Power: When several computers are on a network, a computational grid can be created and the idle computing power of each computer can be leveraged by a computer on the network.

Reliability: Combined resource availability on the network brings reliability. For example, if one terminal fails on the network, a user can use another terminal and complete one's work.

User Communication: A network enables communication among people through instant messaging, email and video-conferencing.

4.6 System Software

System software is computer software developed to run on the computer hardware and enables the application software to run on a computer. System software is often called as Operating System or OS.

Operating system is a very important software in every computer as it manages and interfaces with hardware resources, peripherals and application software. Availability of an Operating System is a prerequisite for any application software to run on a computer.

Examples of Operating Systems:

- Windows 7 or Windows 8
- MacOS
- Linux
- Unix

All the above are examples of an OS.

Activities Performed by an Operating System

Performing Hardware Functions: An operating system interfaces with hardware and communicates the command for the functioning of the hardware.

User Interfaces: Once the system is powered on, the OS provides the interface for a user to use the computer for computing activities.

Hardware-Independent: Operating systems such as Windows, Linux are hardware-independent in the sense that these OS can be installed in any personal computer, whether it is branded or individually assembled. However, **MacOS is not hardware-independent.** MacOS can be installed only in Apple computers.

Providing Application Programming Interface (API): The OS provides a feature called API ro the programmer without resorting to coding for getting the work done with hardware.

Memory Management: The OS handles the memory management activity in a computer. Through a suitable configuration, the OS also helps to create virtual memory by dedicating a portion of the hardware to the computer to supplement the RAM activities.

Task Management: An OS manages all the tasks of a computer by allocating resources for an optimal utilisation of the computer resources. The task management feature of an OS helps a computer to become multitasking.

Networking (if the Capability Exists): Through the LAN card present in a computer, an OS can manage the networking of a computer in a Local Area Network. Even if the LAN card is not present, most of the OSs are capable of connecting to the Internet, provided it is available.

Logical Access and Security: The OS provides logical access and security through user identification and password.

File Management: The OS provides functionality and the feature of file management. For example, Windows Explorer in Windows 7 helps to organise and manage files in the system.

4.7 Application Software

Application softwares are those which function on an operating system software and perform certain useful functions. Application software can also be defined as a collection of programs that addresses the real needs in life. The need could be personal or scientific or business or any other need.

There are different categories of application software:

Office Suite: Office application suite is useful for preparing official documents, preparing certain spreadsheets with calculations and making presentations. Microsoft Office is an example of Office application suite and MS Office consists of MS Word, MS Excel and MS Power Point applications. An application suite consists of several applications bundled together.

Enterprise Application Software: Enterprise application software is useful for handling enterprise sales, purchase, accounting, payroll and similar transactions. Most of the Enterprise Resource Planning softwares are classified as Enterprise Application Software. The examples are SAP, Oracle E-Business, Suite and Microsoft Dynamics.

Engineering Software: This type of software is useful for developing engineering-related work. Examples are Computer Aided Design and Drafting (CADD) software. MATLAB is another example, which helps to simulate some engineering functions through numerical computing.

Enterprise Infrastructure Software: This category of software provides capabilities to enterprise systems. Examples are security systems, antivirus software, etc.

Information Worker Software: These are useful for individual professionals working on projects. For example, MS Visio is a software that helps professionals to prepare a pictorial representation of information such as flow charts. Computer-Assisted Audit Tools (CAAT) is another example of this category.

Content Access Software: This type of software enables the user to view the content of a media file such as audio (wave) files or video files. Examples are media players, Adobe Flash player.

Media Development Software: This category of software is used in media-related work such as editing a video, etc. Examples are Windows Movie Maker and Apple iMovie.

Benefits of Application Software

Computer-Based Solutions to User Needs: Application software provides solutions to almost all the needs of the users—whether personal or professional or business. Especially, with the advent of smartphone-based mobile applications, the users have flooded with millions of applications to choose from.

Less Threat from Virus: When a custom application is built or a packaged application implemented, there is less of a threat from a virus as against the use of applications hosted on the web.

Regular update: Licensed applications are provided with periodical updates by the software vendors.

Disadvantages of Application Software

Development of Custom Application is Expensive: Developing an application from scratch is very expensive and often requires capital investment. Most of the enterprises have started leveraging packaged software instead of developing custom applications on their own.

Threat of Malware: Even in case of custom applications, if invoked through web access through Internet, there is a possibility of infection of malicious software (popularly known as malware).

4.8 Database Management System (DBMS)

The enterprises, knowing their information needs, are required to accomplish the following objectives:

- Acquiring/capturing information on the fly as business transactions happen
- Organising information in a meaningful way
- Assuring information quality and reliability
- Offering tools to the user to access, use and share information

From the above, a DBMS can be defined as an extremely complex set of software programs that controls the organisation, storage and retrieval of data in a database.

The above objectives are accomplished by enterprises leveraging the Database Management System (DBMS).

Functions of Databases

DBMS software performs the following functions:

To Add Files to Database: As and when the need arises, the enterprise users can create a new database file.

To Remove/Delete Files: Often enterprises experience several unused database files due to changes in the business dynamics. A file created several years ago for a given purpose may no longer be needed as business needs would have changed. The DBMS software provides the ability to the enterprise's user to remove or delete the unwanted file. Generally, before deleting a file, it is best to save a copy of the deleted file elsewhere.

To Add Records to the Files: As business transactions happen, the transaction data gets captured on the fly and is stored in the DBMS by adding additional records to the database files.

To Modify/Edit Records in the Existing Database: Often, users experience the need for changing the value of a data entered in the database due to a variety of reasons. The DBMS software facilitates the modification and editing of data entered in the database. Generally, this functionality is provided to the users through an application program, instead of directly making changes in the database. Also, as part of internal control, not all users have authorisation to change/edit the data in the database.

To Delete Data from the Database: Similar to editing or modifying data in the database, deleting a record from the database is another feature offered by DBMS. Though the DBMS software provides this functionality, the best practice is to not to delete the data; instead, to have a field in the file, which indicates that the data is inactive (or considered as deleted). This is needed from the internal control perspective.

To Retrieve/View the Data: the DBMS software enables the user to retrieve and view data. Again, in practice this function is carried out in a structured manner through an application program, instead of users directly accessing the database. Also, this functionality is provided to the users based on their need to view and is controlled through role-based authorisation.

Examples of Database/RDBMS

Here are some examples of database products or RDBM:

- Oracle databases
- IBM DB2/UDB
- Microsoft SQL Server
- My SQL Server (Open Source)

Data and Databases

Database: A database can be defined as a file or a collection of files

Files or Table: A file contains a collection of records. A file is also called as table.

Record: A record consists of fields; each filed represents information relating to the record.

Fields: A field is a collection of data. A particular field in a file will contain same information for all records, but will have different values for each record. For example, in a student database, the field name of the student will have same information—i.e. student name—in all records. But, the value in the field name of the student in each record will vary—i.e. for record 1 it may be Aditya; for record 2, it may be Bharat and for record 3 it may be Chandra.

Data: Data are values of different fields in a record. The data is stored in bit-and-byte format in a database.

The diagrams provided below will explain the above concepts.

Diagram 4.5 Hierarchy of Database Elements

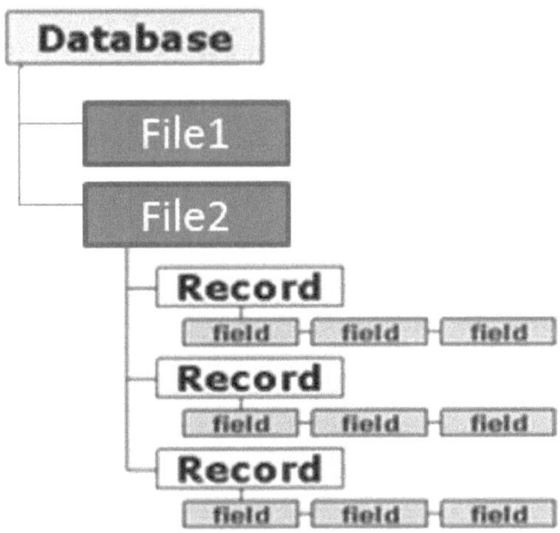

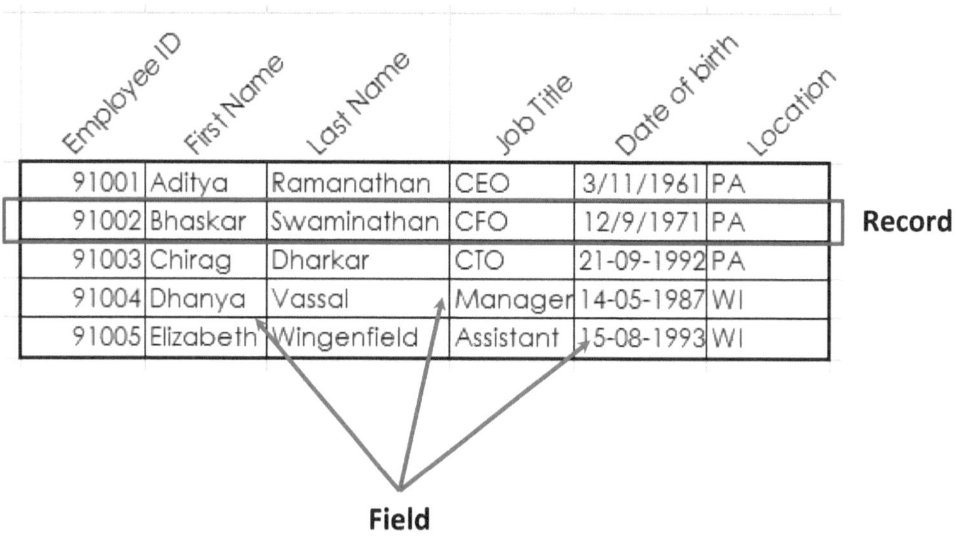

Diagram 4.6 Differences between Record and Field

Database Models

Based on needs, there are several ways in which a database can be modelled.

A. Hierarchical model

Under this model, the records are organised in a database based on hierarchy. It is like an inverted tree pattern. All records in the database are called nodes and each node is related to the other based on the parent–child relationship. The highest-level node is called root.

Under the hierarchical model, there can be only one parent record for a child record but a parent record can have multiple children records.

The diagram given below will visually explain the hierarchical model database.

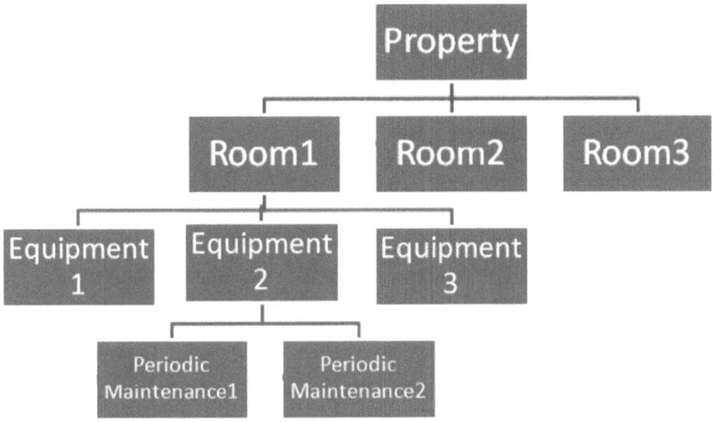

Diagram 4.7 Hierarchical Database

B. Network Database model

In contrast to the hierarchical model, under a network model, a node can branch out from multiple higher nodes and branch into multiple lower nodes.

Consider the diagram below:

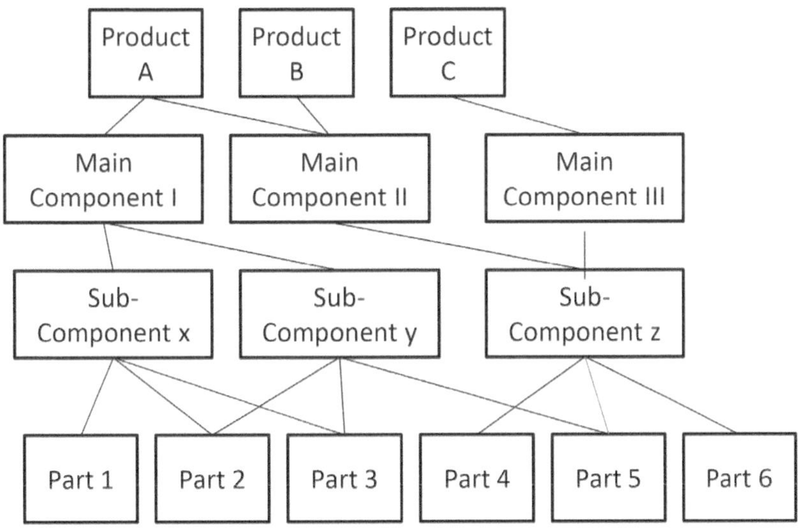

Diagram 4.8 Network Database Model

In this example, Main Component II is branching out from both Product A and Product B; similarly, Main Component I is branching out to both Subcomponent x and Subcomponent y.

In the same way, Subcomponent x branches out to Parts 1,2 and 3 and Parts 2 and 3 are again branching out from Subcomponent y.

C. Relational Database Management System model

The relational database management system (RDBMS) model was invented by IBM. In this case, the database tables are separately created and interrelation between the databases is established through a common key. Consider the following example:

A relational database allows the definition of data and its structures, storage and retrieval operations and integrity constraints that can be organised in a table structure.

There are four terms to be understood in an RDBMS as below:

Relations:Relations are another name for tables with columns and rows.

Attributes: Attributes are named columns

Tuple: A row in a table is called Tuple

Domain: A domain is a collection of potential values for a column. Thus, a domain is a set of values which a column is allowed to take. This is designed mainly for data integrity.

Here is an example of a table in a relational model:

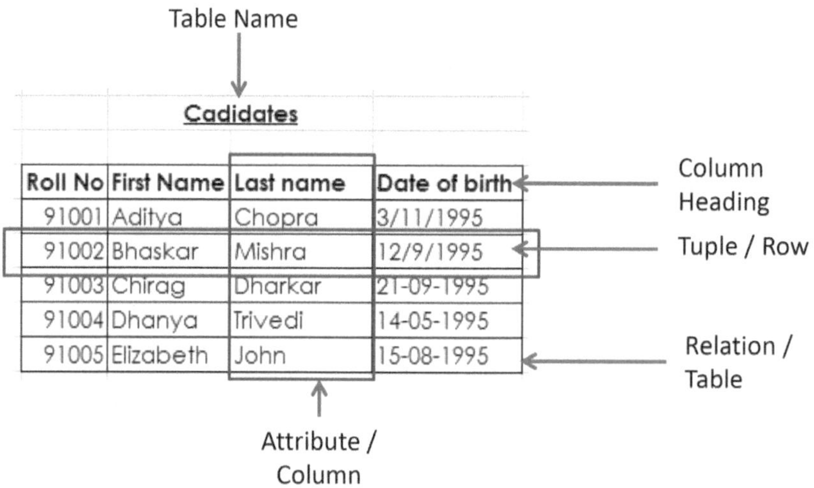

Diagram 4.9 Relational Database Model Elements

In a relational database, multiple databases exist and at least one column common to two or more tables and values is also similar. Thus, two different records in the same or different tables imply a relationship based on the common column (attribute/field).

Consider the example given below:

Cadidates

Roll No	First Name	Last name	Date of birth
91001	Aditya	Chopra	3/11/1995
91002	Bhaskar	Mishra	12/9/1995
91003	Chirag	Dharkar	21-09-1995
91004	Dhanya	Trivedi	14-05-1995
91005	Elizabeth	John	15-08-1995

Fee Payment

Roll No	Amount	Date of Payment
91001	$3,000	8-Aug-14
91002	$2,500	10-Aug-14
91003	$2,750	12-Aug-14
91004	$3,250	16-Aug-14
91005	$2,250	11-Aug-14

Common Attribute / Field

Diagram 4.10 Relationship in RDBMS databases

In the above example, Roll No. is the common attribute between tables and the relationship is established based on this attribute. As one can interpret, there are two payments received from two students with roll nos. S1 and S4.

Thus, in a relational database management model, the relationship between the table and records can be assigned using the cardinality principle1:1, 1: M, N: 1 or M: N.

D. Object-Oriented Database model

Object-oriented programming has helped to evolve the concept of Object-Oriented Database Model (OODB). In the case of OODB, the data is stored in an object instead of a table. In case of OODBMS, the object database is called 'class' as against 'relation' in the case of RDBMS. Similarly, we will have object instances instead of Tuples/rows in RDBMS and instead of column in RDBMS, we have class attribute. A class may have subclasses sharing common attributes between class and subclasses.

Consider the following example:

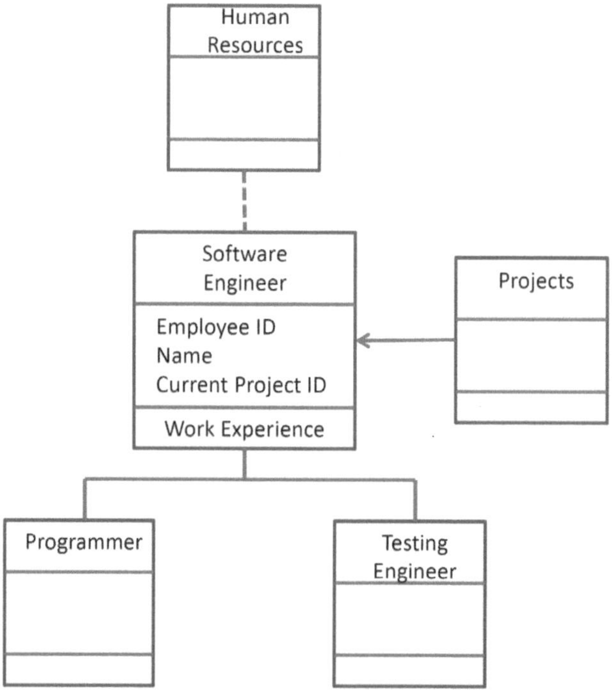

Diagram 4.11 Object-Oriented Database Model

In the above example, 'Human Resources, 'Projects', 'Programmer', 'Test Engineer' are a separate class each.

'Software Engineer' is part of the structure of the 'Human Resources' class and 'Projects' will inherit the attributes of the 'Software Engineer' based on the instances of the 'Employee ID' in the team. Similarly, when 'Projects' are filled with instances of 'Employee ID', the 'Human Resource' class will inherit the value for the 'Current Project ID'. The class 'Human Resource' has two subclasses i.e. 'Programmer' and 'Test Engineer' and both of these subclasses will have 'Employee ID' as a common attribute.

The object classes are related to each other through class attributes and the attribute values are enabled for inheritance under object-oriented programming.

Object-oriented database model is very helpful when the programming is object-oriented as this reduces a significant amount of code to be written by the programmer.

Of the models discussed above, RDBMS and OODB are currently in use. Hierarchical and network database models have become obsolete but are still in use wherever mainframe computers are in usage.

Advantages of databases are as Below:

Data-Sharing Enabled: DBMS helps in the sharing of data. For example, if the marketing and sales department creates a new customer in customer master, the same data is shared with the accounts department in preparing the receivable sub-ledger with the same customer no. as in the customer master.

Minimising Data Redundancy: In functional automation, the respective functional teams were maintaining their own respective redundant data with different structures and attributes. With the RDBMS enabling data sharing, redundant data maintenance is minimised.

Ensuring Data Integrity: The domain feature of RDBMS helps the user to select a value for a field from the collections of the values maintained in the system; with such features, errors are minimised absolutely and data integrity is enhanced.

Program and File Consistency: Data sharing and data integrity offer the facility to use the database attributes consistently across all programs and all files.

User-Friendly: Some of the functions such as "Stored Procedures" make the RDBMS user-friendly/programmer-friendly.

Improved Security: The functionality of authorisation in the RDBMS helps to secure the data through restricted access.

Achieving Program/Data Independent: RDBMS helps to make the data independent of program. Thus, a database may be used by more than one program, without changing the data and manipulation of the data—if any—to be done can be done in the program.

Faster Application Development: When the database modelling is done independently and is effectively designed, it helps to develop the application development faster.

Disadvantages of DBMS are

Cost: RDBMS databases are very expensive and call for a capital investment.

Security: Irrespective of the establishment of a programming-based control, the RDBMS can be hacked from the back-end and data can be retrieved. Security is a major threat where the database contains sensitive and personal sensitive information stored in an RDBMS.

4.9 Importance of IT in Auditing

Extensive application of IT in business made it gain a high importance in auditing. IT is important for the following reasons:

1. IT is all-pervasive in a business and IT applications are at all levels of business from the transaction level to the strategic level.
2. From ideation to value creation, IT is applied in business and audit. It is a function of assessing the value of a business, expressed in the financial statements.
3. Competitive environment and some regulatory requirements demand IT application and innovation.
4. IT application in business is needed for creating a better customer experience.
5. More and more businesses, professionals and individuals are embracing IT and therefore every business is compelled to apply IT in business.
6. Accounting and audit professionals leverage IT in their professional activities.
7. Accounting and audit professionals have started playing key roles in technology advisory to business.
8. In order to perform accounting and auditing, it is not only sufficient to be 'business oriented' but there is also a need to be 'technology oriented'.

Auditing in an IT environment

Although an extensive application of IT in business has made business processes and dataflow more complex in an automated environment, audit objectives of the traditionally managed business and modern business management through IT applications remain the same.

The common audit objectives are

1. **Existence**: Existence of transactions is a prerequisite for values recorded in the books of accounts; Verify assets, liabilities for their existence and ensure it is real.
2. **Authorisation:** Verify and ensure that all transactions have occurred with management intent and approval and that the competent authority has approved the transactions.
3. **Cut-off:** Verify and ensure that all transactions until and up to the cut-off date have been considered and recorded; All transactions for a given accounting period are recorded.
4. **Compliance:** Verify and ensure that the business has complied with all the regulatory requirements; Accounting records reflect the generally accepted accounting practices and principles and also the enterprise policies and procedures.
5. **Operational:** Verify and confirm that business operations are effective and efficient.

6. Recommendations to management for operational improvement.
7. Recommendations for improving internal controls in business processes.
8. Reviewing audit trial of transactions.

In addition to the above common objectives, auditing in an IT environment calls for certain special activities as detailed below:

A. **Study Technical Aspects:** There is a need for studying and understanding the technical environment of the business and for assessing the adequacy of controls and possibility of exposure to risks. Technical aspects include the understanding of hardware, software, applications, operating systems, databases, network and security systems deployed in the environment.
B. **Using Unique Audit Techniques:** An auditing IT environment calls for the use of a certain unique technique, which starts with the understanding of the technical environment. Besides, a reviewing of system documents and an interviewing of technical specialists are needed. Also, as part of the audit procedure, evaluating and assessing the controls are needed.
C. **Audit Software Usage:** Finally, using audit software to scan through the technical environment to spot the vulnerability and to verify whether what is being stated by the business is truly reflecting in the environment.

IT Risks and Issues

"Higher the technology, greater the need for control" is appropriate for the implementation of emerging information technologies in business. If anything vertically takes off, without adequate control, it will result in a vertical landing. In the early part of this millennium, among other reasons for business failure, an inadequacy of control over the IT system was one of the reasons for failure of Enron and Worldcom and also the dissolution of the public accounting firm Arthur Andersen. Consequent to the Enron and Worldcom debacles, the government in the United States passed a legislation called the Sarbanes Oxley Act, which expects both the management (CEO and CFO) and the auditor to certify that the financial statements represent the true view of the financial status of the enterprise and that there are adequate internal controls over the system and processes in the enterprise. While evaluating internal controls and assessing their adequacy is always part of audit, this emphasis has been explicitly made by regulatory authorities in an integrated IT environment where automated data flow happens between enterprise systems.

In essence, it is the responsibility of an auditor to question the system

- "What type of processes, the enterprise uses?"
    - to identify the business and technology risks
    - Identify risks, when change in technology occurs (new types of services or change of device)
- "What are the risks faced by the enterprise?"
    - What precautions are taken by the enterprise, when it makes changes to the critical system, to prevent the failure of the system due to change.

Control on the Information System

- As part of audit, there is a need for an auditor to evaluate the enterprise control of its information system. This is critical because the information system is considered as the digital nervous system of an enterprise and it impact is multifaceted as detailed below:
- Strategy and direction
- General organisation and management
- Access to IT resources—data and programs
- System development methodology and change control
- Operation procedure
- System programming and technical support function
- Quality assurance and testing procedures
- Physical access control
- Business Continuity Plan (BCP) and disaster recovery plan
- Network and communication control
- Database administration
- Protective and detective mechanism against internal and external threats

Unique Aspects of Auditing in an IT Environment

Unlike traditional audit, in case of auditing a business operating in an automated IT environment, the auditing involves certain unique aspects.

Auditing the IT-related aspects includes:

Hardware: This involves auditing servers, storage—every electronic device—such as router, switches, network, etc.

Software: This involves auditing all applications software, operating system, network-system software, database systems, security software, etc.

Procedure: This is about reviewing the standard operating procedures relating to how the IT systems are working within the enterprise and also how the data communication happens with entities outside the enterprise and in reviewing other internal controls.

People: This is about reviewing who has been entrusted with what responsibilities, whether the principle of segregation of duties has been implemented, etc.

The above four elements operate together to convert the data generated by the business transactions into meaningful information for the business.

The key in an IT environment audit is to (a) to know the audit methodology (b) to plan the audit implementing the appropriate methodology, perform and complete the same.

Impact of IT Risks on Controls

The combined power of information and internet technologies have offered enormous benefits to businesses. As enterprises have started conducting business through Internet and communication technologies in different parts of the world, the volume of data is on the increase. Simultaneously, there is an increase in the risks also.

Consider the following risks:

1. **Ready to Access to Terminal**: Accessing the enterprise's IT environment from an open public system such as an Internet cafe through an Internet browser poses tremendous risk.
2. **Online Processing through Public Internet Cloud** posses risk as strangers and hackers can intercept and access data.
3. However much control is set up in the processing of an online transaction, **perpetrators** innovate newer ways to intercept control and **hack the system.**
4. Enterprises often **fail to recognise the threats.**

The above listed risks impact traditional controls such as

a. **Centralised Data Entry**: Centralised secured data entry processing is disappearing as more and more self-service models emerge such as the customer, vendor, employee self-service portals, etc.

b. **Physical Custody of Document** no longer exists: For example, in the past in order to affect transfer of money from a bank account, one needed to provide a written requisition to the bank and such physical document in custody of the bank became the basis for transaction. With Internet banking, there is no such physical document. If a transaction is successfully completed by providing all credentials of the account holder while logging in and responding to other transaction controls, the transaction is deemed to have been executed by the account holder.

c. **Transfer of Responsibility**: Through a single command over the Internet, transactions can be completed in all respects. For example, in case of Internet banking, the responsibility of bank transactions have been shifted from the bank to individual account holders.

d. **Decline in Accountability of Traditional Functions**: Again taking the example of bank transactions, in the past the transactions were conducted in a branch, hence branch was held accountable and responsible. In the modern age of banking where the processing is carried out by a centralised back office, there is a decline in responsibility of the branch and the customer hardly has any direct access to the centralised back office.

Concerns of Auditors

Increased automation through IT coupled with decline in traditional controls poses the following concerns to the auditors to:

A. Develop a new approach with a newer criteria in auditing a computerised information system.

B. Develop tailored testing techniques.

C. Use computers in performing certain audit functions such as extracting changes made to systems; using system audit software.

Section 5

Network, Internet and Telecommunication

5.1 Background

Unlike the past, the telecommunication facilities now not just offer voice communication alone but also enable data communication. This has enabled enterprises to leverage telecommunications extensively for various business benefits. There are four distinct ways in which the enterprises leverage telecommunication networks.

Interconnection
In the past, the multinational corporations used to have their presence in multiple countries performing their all operations locally and selling in the local markets. For example, if a multinational company was setting up operations in the 1970s, the company used to source the material locally—except the critical material, which could not be locally procured—engaging local labour and selling the products in the local market. In the past, multinational corporations' business operations were more in the form of financial and technical investment in different parts of the world. Now, after commercialisation of the Internet, the concept of multinational corporations has come to mean global enterprises. The meaning of global enterprise is an enterprise having its presence in different geographies and all the entities in different geographies operating as a single entity. The main change is that the enterprise performs its functional activities in a single global location for the entire global operations, based on what is best offered by that location. For example, enterprises have set up their research and development facilities in the US and Europe; their back office operations in South and South East Asia in countries like India, Philippines and manufacturing in the Far Eastern countries like China and Korea. This is possible due to the Internet, which is offered as a service by the telecommunication networks.

Instrumented
Enterprises have started leveraging the technology advancement for their benefit. For example, technologies such as video conferencing and tele-presence have enabled global enterprises to have an effective communication an effective communication several thousand miles. Similarly, remotely serving the entire global operations is another feature of off-shore global delivery centres of IT-enabled services. Primarily, Internet services offered through telecommunication networks is the reason for the enterprise getting instrumented.

Intelligent
The entire global society is becoming more intelligent in the Internet era. So as with the global enterprises; the information with one individual in this world can be made available to the

rest of the population of the world. Similarly, enterprises have started gaining knowledge and intelligence on each of their functions through the information available on the Internet. With Internet, enterprises are well-informed and executives are enabled to take information-based decisions.

Inter-enterprise Cooperation

Internet enables an enterprise to come together and collaborate and evolve a new business model for the benefit of all. For example, the post-Internet era has experienced the e-Commerce boom. e-Commerce is a new business model, where the e-Commerce portal owner is able to showcase and sell the products of different brand owners. In this process, there is a collaboration happening among different businesses—product brand owner and its retail chain, logistics service provider, payment system service provider, banks and the e-Commerce portal owner.

Besides e-Commerce, Internet has offered an enormous opportunity for the enterprises to automatically transfer data between them, thus minimising the time delay by removing geographical barriers and without much of the human error. Again, this is possible because of the Internet service being offered by the telecommunication networks.

5.2 Network Architecture

Tier means a distinct part of or a group of entities. In network architecture, a tier means a distinct part of hardware or software.

Multi-tier architecture actually has two classifications:

a. Two-tier network architecture
b. n-tier network architecture

By default, every single computer system is part of one-tier architecture. It is also called as single-tier network architecture.

1. One-Tier Architecture or Single-Tier Architecture

In this case a single machine consists of user interface, applications and database. This type of architecture is often used by small businesses.

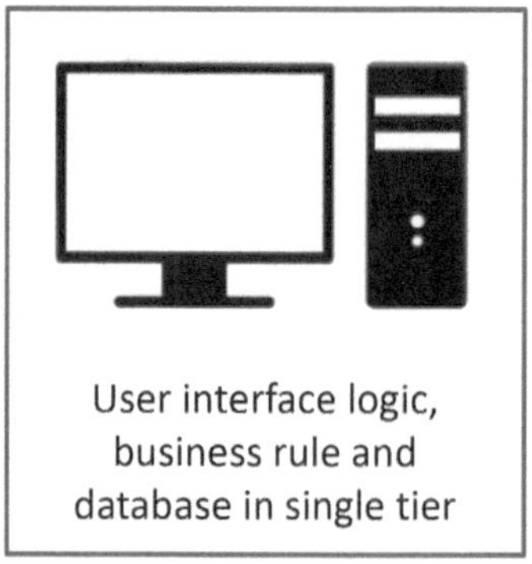

Diagram 5.1 One-Tier Architecture

The advantage of a single-tier architecture is less investment; hence, it is cost effective. Only one system and a set of application software and a database are required.

The disadvantage is that only one user can use the system at any point in time.

2. Two-Tier Architecture

Two-tier architecture is a true representation of the client-server architecture, where client machines will have user interface and business applications and the server will host the database.

A diagrammatic expression of a two-tier architecture is given below:

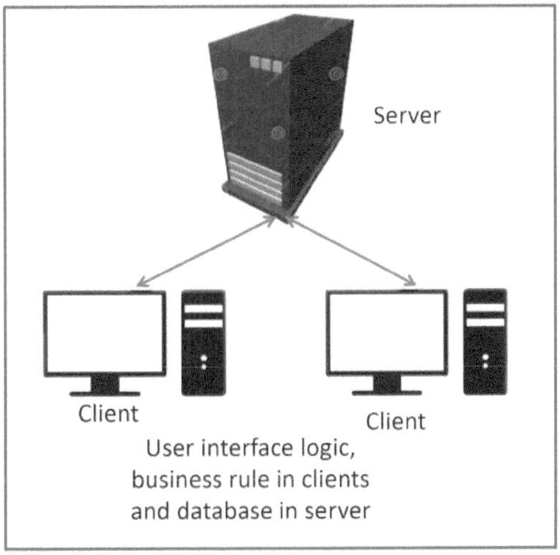

Diagram 5.2 Two-Tier Architecture

Advantages:

a. As it is a true representation of the client-server architecture, a number of clients can be set up according to the number of the users requiring the computing resources.
b. A server that hosts a database is shared by all users.
c. Compared to an n-tier architecture, this is relatively easier to set up.

Disadvantages:

a. Not scalable if too many users are added, unless an additional server resource is added.
b. It is not a flexible database choice for the programmers.
c. Application maintenance is tedious as multiple copies of the same application will exist in multiple clients and therefore, if any upgrade of any application is to be effected, it has to be carried out as many times as there are the client systems holding it.

3. "n-Tier" Architecture

Under an-tier architecture, the client holds only the user-interface components. Applications are set in a separate server as a second layer to the network and databases are set up in another third layer in a separate server. If only one application server is hosted, then this is called as 3-tier network architecture. However, if multiple applications servers are hosted, this is called n-tier architecture.

Consider the diagrammatic representation of an n-tier architecture (3-tier architecture) given below:

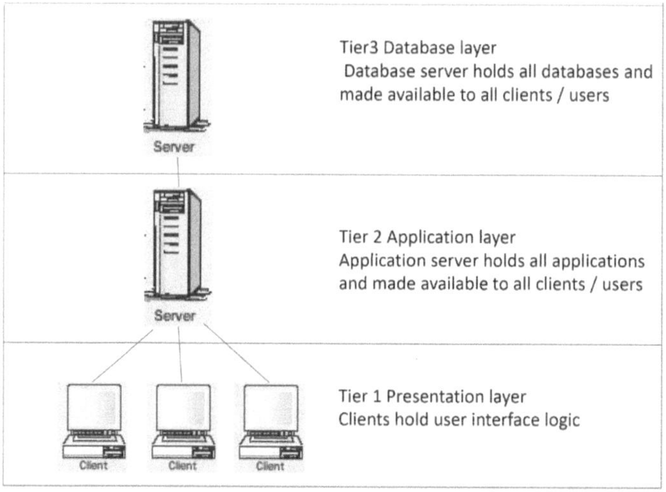

Diagram 5.3 n-Tier Architecture

Advantages:

a. Clear compartmentalisation and segregation of layers/tiers based on functionality

b. As applications and databases are distributed to different layers, load gets distributed among resources of the different tiers. Further, server resources can be optimised through load-balancing techniques.

c. Updates and upgrades of applications and databases can be maintained very easily as single updates in a server to be carried out.

Disadvantages:

a. More the tiers, more the network management workload increases

b. n-tier architecture calls for high investment as more resources have to be set up

c. If any one layer of application tier or database tier fails, it makes the entire network go down.

d. Higher maintenance costs as several resources and several tiers have to be maintained.

Networks are of different types. Broadly there are three types of networks as below:

1. Area-coverage-based
2. Function-based
3. Ownership-based

Further classifications of different networks under each of the above and the high-level descriptions are provided below:

Area-Coverage Based		
Local Area Network (LAN) A LAN network is a group of computers connected together within a same entity—a building or a campus—for a high speed data transfer. It is relatively less expensive.	**Wide Area Network (WAN)** A WAN is not bound by a geographical location. For example, a global corporation may have a WAN covering multiple geographical areas	**Metropolitan Network (MAN)** A MAN network is usually a network that covers an entire city. For example, telephony network in a city.

Function-Based		
Client Server It is a kind of network where a server is connected to different clients. The server handles the workload of different clients.	**Peer-to-Peer (P2P)** This is a network where all the participants in the network are peers to each other. There is no client or server. The peers equally provide or avail services.	**Multi-Tier** It is a network architecture model where the server side is organised in two tiers; the first being the application tier and the second being the database tier.

Ownership-Based		
Public Network A public network is open to the public to freely participate in. For example, open Internet cloud is a public network.	**Private Network** A private network is restricted for participation. The participants need to meet some criteria and are often authenticated with user identification and a password to log on to the private network. Most of the network of enterprises is private network.	**Virtually Private Network** Virtually private network is creation of a private network in the public Internet cloud. It is like constructing a virtual tunnel on the public Internet cloud and then making it private. Again users are authenticated before participation in a VPN. Data transmitted over VPN is encrypted and decrypted for security.

Area-Coverage-Based Networks

A. Local Area Network (LAN)

As the name suggests, a local area network connects devices of a local area—say, a building or campus. This type of network is extensively leveraged by business enterprises.

Characteristics of LAN: A LAN network may use twisted pair or coaxial cable or fibre-optic cable as media. A LAN carries both voice and data over the network. A device can connect over network through an unbound media—say, a Wi-Fi router.

For a personal computer to connect to the network, a network interface card or a Wi-Fi receiver should be present.

A LAN, generally, is set up with a high-powered computer with a large disk capacity as server with a network operating system such as Novell Netware or Windows NT/Windows Server.

End-users in a LAN environment are called a work group and can communicate with each other over the network and share resources.

A LAN environment enables one to have a distributed computing environment and a powerful computing facility can be a part of LAN and can be shared with all the end-users.

Features of LAN

1. **Security**: A LAN environment can be made secure allowing the users to participate in the network with user authentication and password. Besides, access can be restricted for selected folders and files.
2. **Thin Client and Powerful Server**: A LAN environment can be operated with thin clients and powerful server, whereby the clients are inexpensive, and a powerful, centralised high capacity computer called 'server' made available as common resource for all users.
3. **Distributed Computing:** A LAN environment offers facility to share resources, thus enabling distributed computing.
4. **Mail and Instant Messaging:** Emailing and instant messaging functionality can be made available in a LAN environment for the benefit of the participating users.
5. **Multiple Advantages:** An enterprise LAN offers multiple advantages—a reduced cost of the computing environment, centralised data management, backup and maintenance, A single copy of application software used by multiple users helps an easy upgrade of the software, centrally.

B. Wide Area Network (WAN)

Telephonic network is the best example of WAN. An enterprise may have a wide area network, when it spreads across different locations. Tier-1 Internet service providers have their own WAN that spreads across different continents, connected through a fibre-optic cable running under the ocean or through a satellite.

C. Metropolitan Network (MAN)

A metropolitan network is a single network that connects all the devices in a city. Under the concept of smart city, most of the cities are nowadays Wi-Fi-enabled and those are good examples of MAN.

Function-Based Networks

A. Client Server Network

Client server network is a very common network architecture adopted in any business enterprise. A client server environment enables a distributed computing and interconnected system environment in an enterprise. It enables an enterprise to optimally distribute the computing resources. Clients and server are the components of a client-server computing environment.

Client

A client is a component of client-server network architecture, where the user performs computing activities. A client is a workstation that offers presentation services to the user and from the server it also provides database services to the user. Based on the nature of activities that can be performed, clients can be classified as (1) Thin client (2) Fat client or Thick client and (3) Hybrid client.

Thin client

A thin client generally will not have any major computing resources. It enables the user to connect to the central server and provides computing power to the user by connecting the user to the central server. Thus, a thin client acts as a presentation layer in the client-server network architecture.

Fat client or Thick client

A fat or thick client is capable of performing a bulk of data processing activities by itself. Therefore, unlike thin client, a thick client does not rely on the server resources for performing the computing activities. A thick client is common in a two-tier network architecture, where the server hosts only the database.

Hybrid client

Hybrid client is one that has the features of both thin client and thick client. Like a thick client, it has its own computing resources, yet for storage it uses server like a thin client. Thin client is well suited for gaming consoles.

Server

A server is a one or more multi-user processor with high power processing capability and with a large storage capacity. In a client-server network architecture a server is connected to the clients and functions as a sharable resource.

How a does a client server network work?

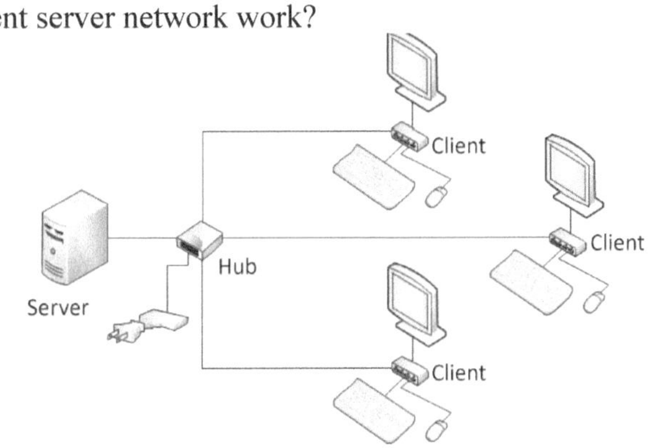

Diagram 5.4: Client Server Network

Consider the above diagram:

1. The above diagram is an example of the client server network.
2. The network contains three components: Server, hub and clients.
3. Servers are powerful computers capable of hosting applications and databases.
4. The server ensures security by validating the user-identification credentials, when someone attempts to log on to the network.
5. The clients are also called workstations, where the end-users perform their computing activities using the resources of both client and server.
6. Often the actions (command) are triggered from the client end.
7. A server is capable of servicing multiple clients at the same time.
8. A network operating system is implemented for managing the network. It is also known as 'Middleware'.

Characteristics of a client-server network architecture

1. Service-based architecture: A client server architecture enables a network architecture based on services. For example, 2-tier architecture, n-tier architecture.
2. Sharing of resources: In a client-server network architecture, the sharing of common resources is enabled. For example, a printer in a network can be shared by all users; the server itself a common shared resource.
3. Location identity of server is masked by the network software, when clients are working with the server. This is both an advantage and a disadvantage.
4. Network operating systems are independent of the hardware set-up. For example, any branded or assembled server can install Windows NT or Novell Netware or Linux software.
5. Network as a whole is scalable: More clients can be added and as many servers can be added to make the network more powerful.
6. The integrity of data and application are assured as the applications and data are, generally, stored on the server and centrally maintained. All clients use the same set of data and the same version of the application at any point in time.

Risks with Client Server Architecture

a. When a server goes down all the clients are inoperative—particularly, if the clients are thin clients deriving computing resources from the server.
b. There is an element of time lag in processing data in a client-server environment as the client has to communicate the processing data to the server and the server gets loaded with such requests from several clients simultaneously.

B. Peer-to-Peer Network Architecture

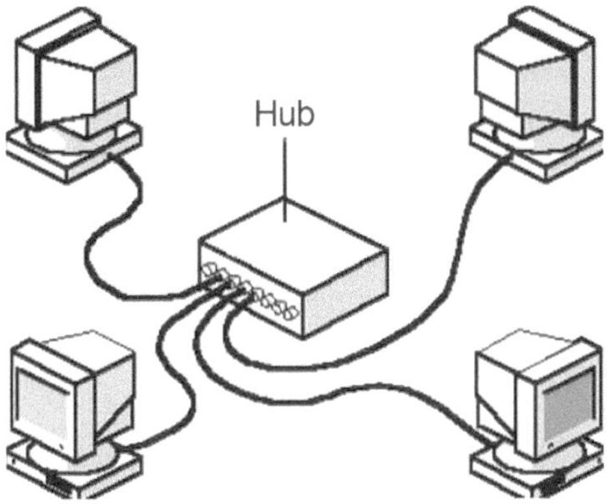

Diagram 5.5 Peer-to-Peer Network

A peer-to-peer network is where two or more computers are connected through a hub or switch. In case of a peer-to-peer architecture, there is no client and no server as all the devices participating in the network are considered peer to each other—hence, the name.

The main objective of a peer-to-peer network is to pool the resources of all the participating devices and share the resources. In addition, file sharing and communication are features of the peer-to-peer network architecture.

In case of a client-server environment, where the client requests for services from the server (more particularly in the case of a thin client environment), it can be said that communication is unidirectional; but, in the case of a peer-to-peer network environment, the communication is bidirectional, where every participating resource can ask for resource from other participating devices and also offer its own resource for sharing with others.

Advantages:

a. Easy to set up
b. Less expensive
c. Even one peer participant device fails, others can continue to be active.

Disadvantages:

a. Scalability is an issue
b. There is no centralised security management as security is with each peer participant separately
c. Even if one peer participant is down, it may impact the others as files to be used may be residing in the peer participant that is not operational.

Ownership-Based Network Classification

A. Public network

A public network is a network for the public and anyone can participate in the sense that participants from the general public network may not be from the same entity—business enterprise or educational institution, etc.

A public network is offered by a private agency such as the telephone service provider.

Internet is an example of the public network.

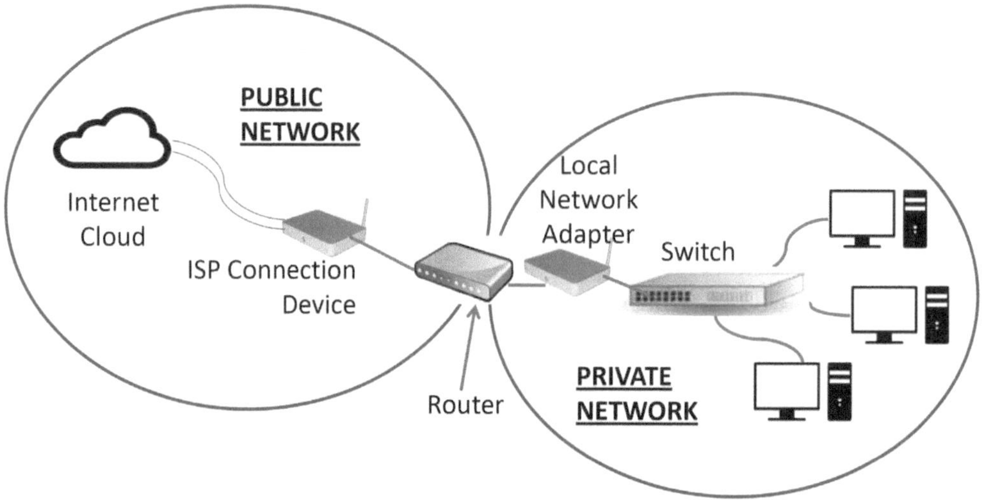

Diagram 5.6 Internet and Intranet

B. Private network

A private network is the one that is not a public network—a private network is not allowed for use by the general public. It is a network exclusively for an entity and participants belong to the same entity. A private network is managed by the entity who owns it.

A participant of a private network can access the public network—Internet—when the private network is connected to the public network, though the general public cannot access the private network.

C. Virtual Private network

As the distributed computing came into practice along with the public network (Internet), there was a strong need felt by the business enterprises that in the process of interconnecting the geographically distributed locations of the entity, they should be connected the Internet and yet achieve privacy in their network operation. Thus was created the Virtual Private Network or simply VPN.

By having a VPN set-up, a user can access a private network through the public Internet.

Basically, the VPN set-up authenticates the user identification credentials and forming a invisible tunnel in the public Internet, allows the user to access and transact on a private network.

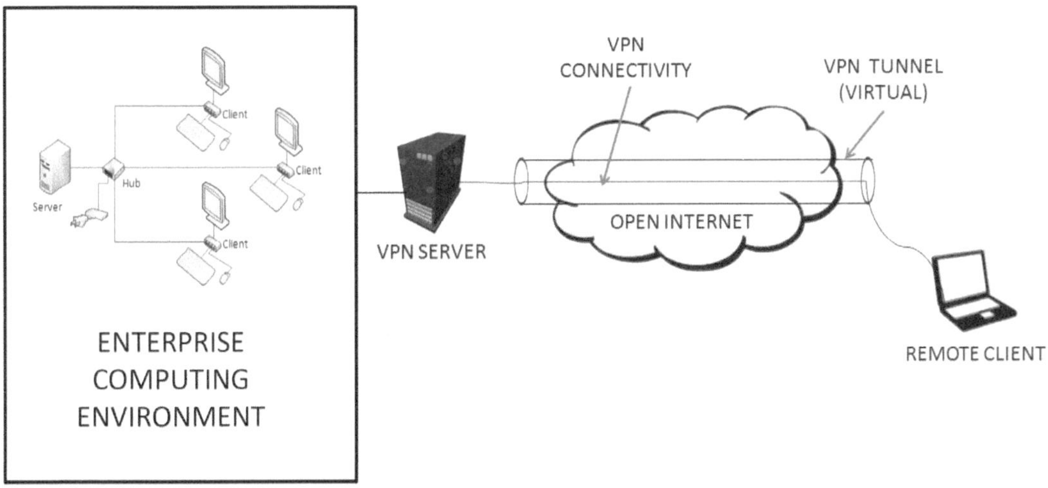

Diagram 5.7 Virtual Private Network

Internet is the real backbone network and a prerequisite for the VPN to function.

5.3 Network Computing

Over the decades, computing models have matured with a client-server network model with an n-tier architecture. These models and architectures, in turn, offered business benefits through intranets, extranets and Internet. A further advancement is the accessing of a private network through the public Internet with the support of VPN; the network computing is extensively in use among business enterprises.

Network computing has helped the enterprises in achieving standards in IT practices. Given the goal of maximising benefits and minimising investments and costs, network computing is common practice among mid-size and large enterprises.

As part of cost optimisation and standardisation, many enterprises have adopted client-server architecture with thin client as option. Such environments offer: (1) users to access the server through browsers (2) there is no storage or memory resources in the thin client environment, hence users can perform computing activities on the central server (3) the server provides the operating system, applets (a client application to access applications in the server) and database resources needed for a user. This is also called 'centralised computing'.

Centralised Computing

As the name suggests, a centralised computing environment has powerful servers that are shared resources for all users. The user accesses the servers or central computer through the terminals that are dumb. The dumb terminal is also called as thin client that has only the feature of connecting to the central computer and run certain applets that invoke the application hosted in the server. All computing activities for all the users are centralised.

The advantages of centralised computing are

(1) Sharing of resources (2) High security (4) Even if one terminal is not functioning, the user can perform computing activities through another terminal (5) The investment is relatively less than that of a decentralised computing environment.

Centralised computing is not without disadvantages. The main disadvantages are

(a) If the server fails, the entire network us unavailable for computing activities (b) computing activities of a business significantly relay upon the effective and efficient centralised administration.

Decentralised Computing

In decentralised computing the resources such as hardware, software and network are allocated to different business operations for conducting the computing activities of the business. Generally, in the decentralised computing environment, certain or all the client devices are capable of independently performing computing activities, as clients will have memory and storage features.

Generally, a collection of decentralised computing environments is networked together and the client derives powers from centralised servers and also leverages its own resources.

The single biggest challenge in a decentralised environment is its maintenance.

Network Topology

Network topology is about the way in which the clients are structured and arranged.

There are a number of ways in which the clients in a client server network can be arranged and connected. The popular ones are Bus network, Star network, Ring network and Mesh network.

A. Bus network

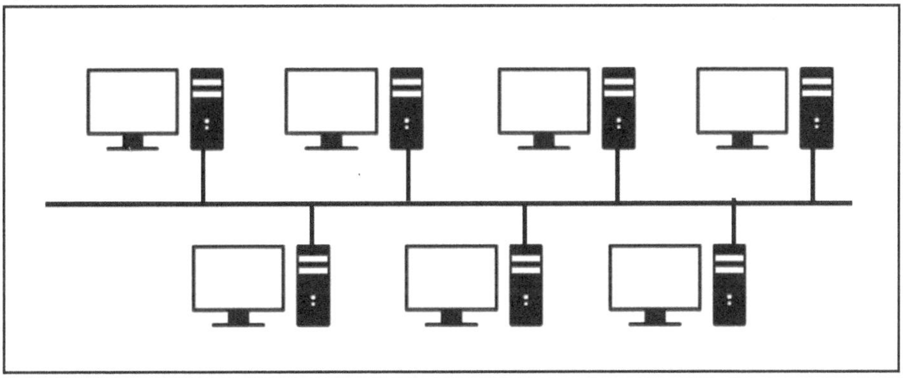

Diagram 5.8 Bus Topology

It is a long single cable—twister-pair wire or fibre-optic cable that runs along and connects all the clients—which is called a bus network. All communication travels along the single cable called a bus. It is a decentralised network.

Advantages

a. There is no host computer or server, hence, the bus network is reliable and easy to understand.
b. If one computer in a bus fails, rest of the computers can continue to function.
c. Consumes less network material as a single cable suffices for setting up the network.
d. Bus is scalable, but with a repeater after a certain length of the bus.

Disadvantages

1. Heavy use of network and higher traffic can make the network slow.
2. As the bus become longer the signal weakens.
3. Very difficult to troubleshoot, if there is a cable cut.
4. If there is a disconnection in the bus cable, the entire network becomes unavailable.

B. Star network

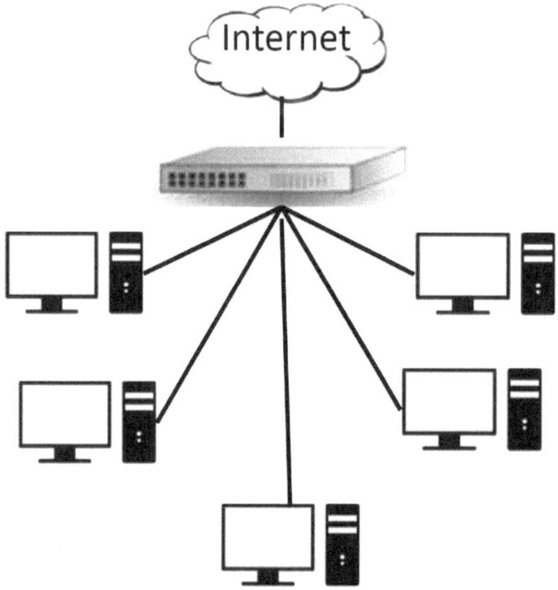

Diagram 5.9 Star Topology

Star network is a popular topology, where all the terminals are connected to a central unit.

The central unit is generally a mainframe computer, which is a host. The host manages the network traffic. It is more suitable for a huge volume of data-processing activities. A decentralised computing that uses a star network creates a hierarchical topology.

Advantages

1. Several terminals can be added to the network.
2. Easier to add and remove terminals from the network.
3. If one terminal fails to function, still work can be carried on with another terminal.
4. Troubleshooting the network is easier with the help of network monitoring tools in the host computer.

Disadvantages

1. The whole network is unavailable, if the host is unavailable for computing activities.
2. Calls for heavy investment in setting up the star network as it will demand more materials.
3. Considered less reliable than a ring network, as there is a heavy reliance on the host server.

C. Ring network

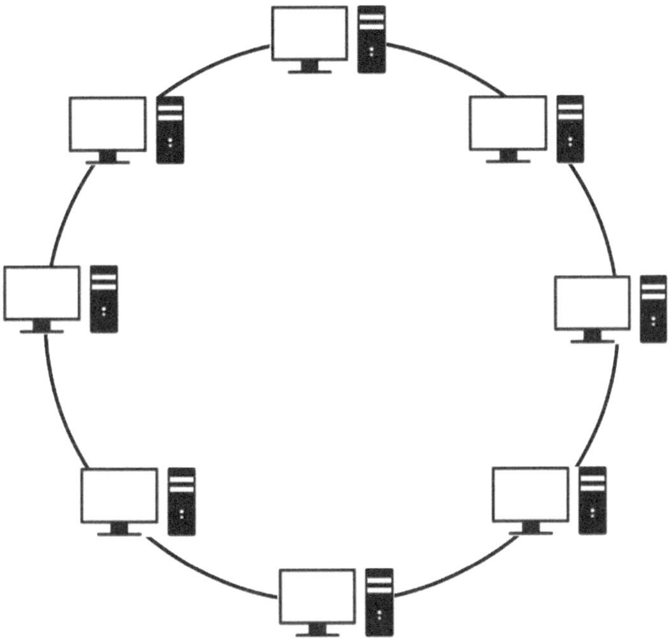

Diagram 5.10 Ring Topology

The ring network is something similar to a bus network but with a difference: the single cable is laid as a loop instead of as a straight line. Like the bus network, the ring network also does not need a host or a server.

Advantages

1. A ring network does not need a host or a server.
2. Each computer in the network can communicate with the other computers along the ring.
3. Even if one computer fails, the network continues to work with the other computers in the network.
4. A ring network is suitable for smaller networks and for high performance.
5. Ring network is expandable and extensible.

Disadvantages

1. Very expensive network as each terminal is a separate computer with computing resources
2. If the cable laid in the form of a ring fails, the entire network will become unavailable
3. Difficult to troubleshoot as there is no host with network management software
4. With every addition and removal of a terminal, the network has to be brought down.

D. Mesh Network

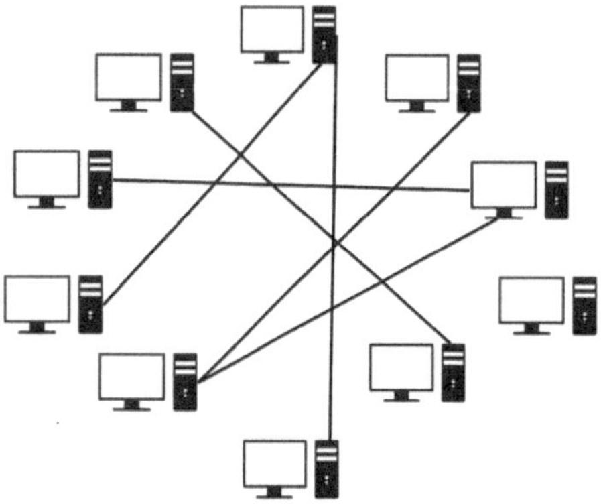

Random connection Mesh Topology
Diagram 5.11

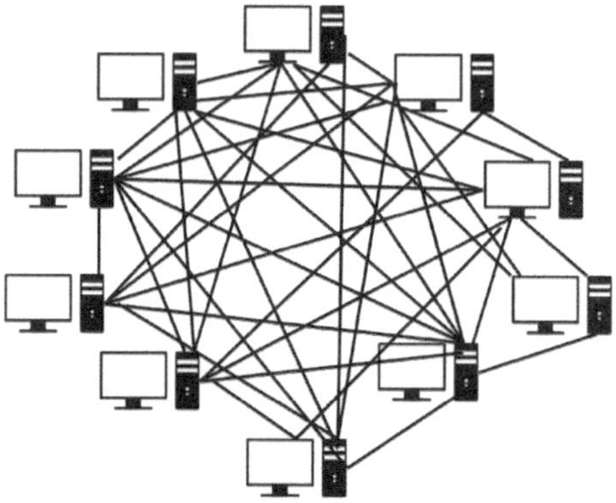

Fully connected Mesh Topology
Diagram 5.12

A mesh network is a topology where the nodes are either randomly connected or fully connected with one another. In a fully connected topology, every node is connected with the rest of the nodes in the network. This topology is highly reliable as there is always an alternative path available in the network, if one path is unavailable due to some reason.

Advantages

1. Highly redundant with alternative paths
2. Easy to diagnose network problems

Disadvantages

1. High investment required
2. In a fully connected network, adding a new node or removing an existing node is a challenge as the network needs to be established or disconnected respectively, with the rest of the nodes in the network.

Transmission Technologies

Data, voice and images travel through network, sent by one system and received by another systems. Technologies around such transmission of data are important to understand. Transmission technologies can be understood from multiple perspectives: (1) Transmission mode: number of bites sent simultaneously (2) Synchronisation between transmitter and the receiver (3) The direction of the exchanges and (4) Various transmission techniques

A. Transmission Mode

Transmission mode in a network data flow refers to the amount of data transmitted simultaneously through the carrier. There are two modes under which the data transmission is carried out (i) Serial transmission and (ii) Parallel transmission.

Serial Transmission

In a serial transmission, bits of each byte are transmitted one after the other, sequentially through a single path. An example of a serial transmission is a mouse communicating to the computer.

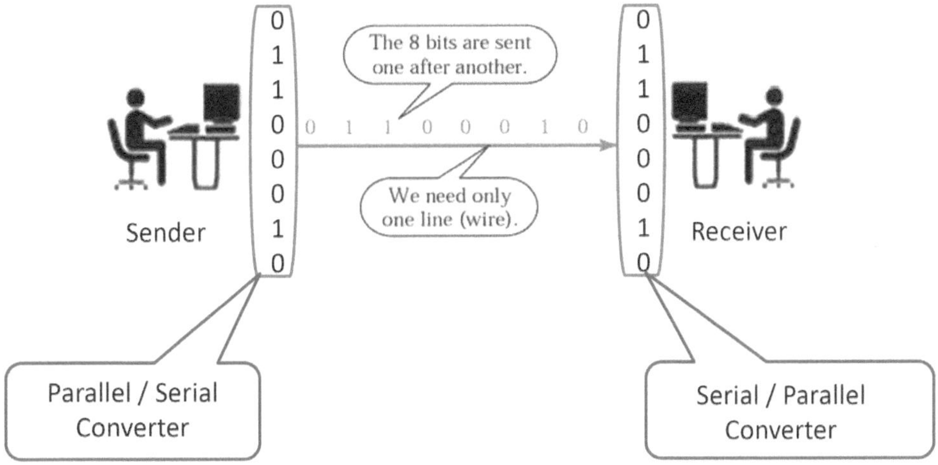

Diagram 5.13 Serial Transmission

Parallel Transmission

In a parallel transmission, all the bits in a byte are transmitted simultaneously. An example of a parallel transmission is Centronic port used in printers.

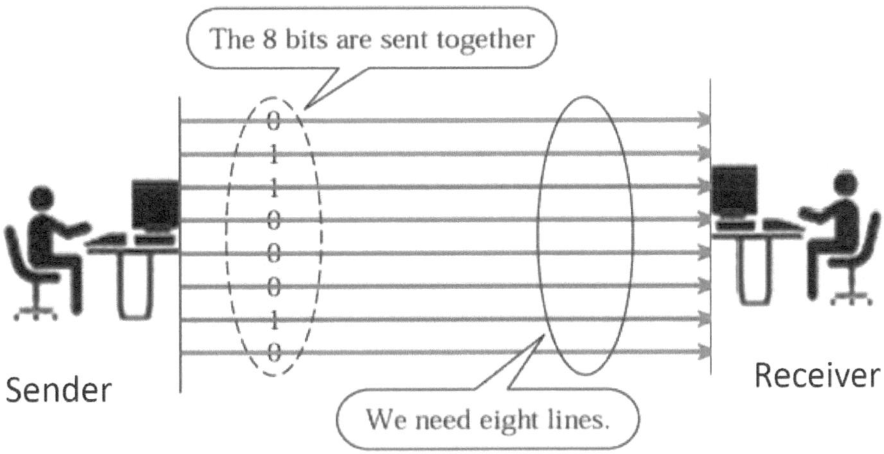

Diagram 5.14 Parallel Transmission

The differences between serial and parallel transmission are listed below:

Differences between Serial and Parallel Transmissions

Serial Transmission	Parallel Transmission
Transmits one bit at a time sequentially	Transmits all the bits of a byte simultaneously
Data transmission through single wire	Data transmitted through 8 wires to transmit 8 bits at a time
Economical way of transmitting data	Expensive transmission
Suitable for long or short distance	Suitable only for a short distance as long distance parallel transmission may cause error in simultaneous transmission of all bits in a byte
Slow in transmission	Relatively faster in transmission

B. Synchronisation between Transmitter and Receiver

There are two types of transmission from the synchronisation perspective:

Asynchronous transmission

It is possible that the clock pace of the sending device and the receiving device may not be of the same speed—therefore, the transmission of data is carried out at irregular intervals. In order to recognise the information, there are the START bit and the STOP bit at the beginning of the information transmission and at the end of the information transmission, respectively. This enables the receiving device to decipher that between the START bit (0) and STOP bit (1), the message bits are to be grouped and decoded. There will be a gap between the data units.

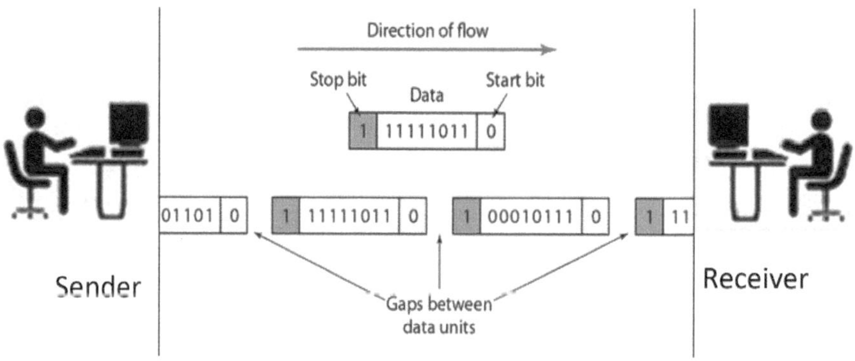

Diagram 5.15 Asynchronous Transmission

Synchronous transmission

Under synchronous transmission, the data is transmitted through a continuous stream with the data signal being accompanied by time signals to synchronise the time clock of both the sender and the receiver. The data is sent in a block at fixed-time intervals.

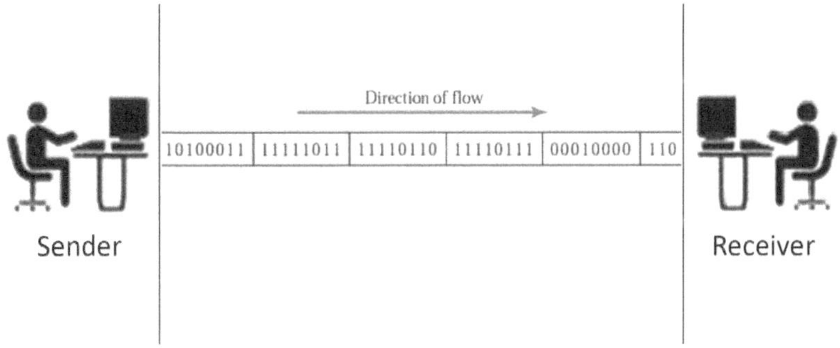

Diagram 5.16 Synchronous Transmission

Differences between synchronous and asynchronous data transmission:

Synchronous	Asynchronous
Message data is sent continuously without start and stop bits	Each message data is prefixed with a start bit and suffixed with a stop bit
Transmission is continuous and faster	Transmission is intermittent and slow as there is a gap between every stop and start
Synchronous devices are expensive	Asynchronous devices are relatively more economical
Possibility of data loss higher	Possibility of data loss lower
Efficient	Effective

C. Transmission based on direction

There are three types of transmissions based on the direction of transmission: (1) Simplex (2) Half Duplex and (3) Full Duplex.

Simplex

In this case, the data transmission happens only in one direction. This type of transmission is set up where the data need not flow in a reverse direction. For example, connection to a printer.

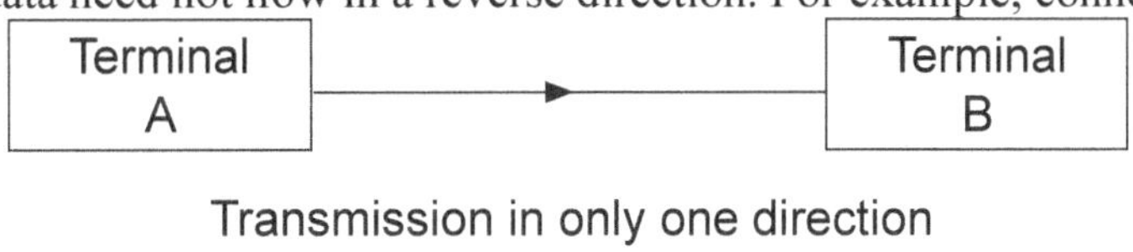

Diagram 5.17 Simplex Transmission

Half Duplex

In case of a half duplex transmission, the transmission happens in both the directions. But, at any point in time, the transmission can happen only in one direction. An example for this is theWalkie-Talkie.

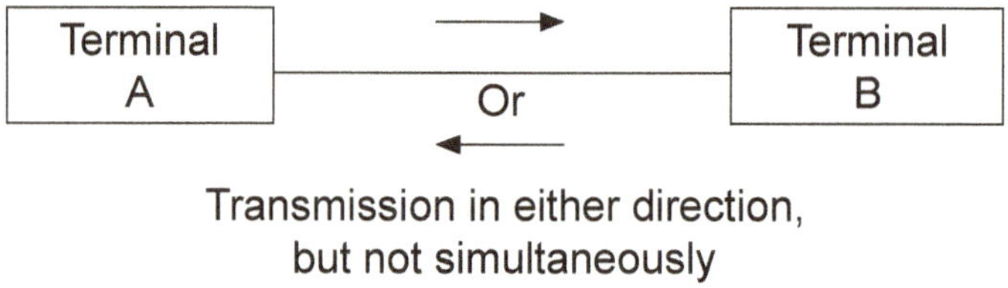

Diagram 5.18 Half Duplex Transmission

Full Duplex

Under a full duplex transmission, the data transmission happens in both the directions without any restriction. An example for a full duplex transmission is mobile phone communication.

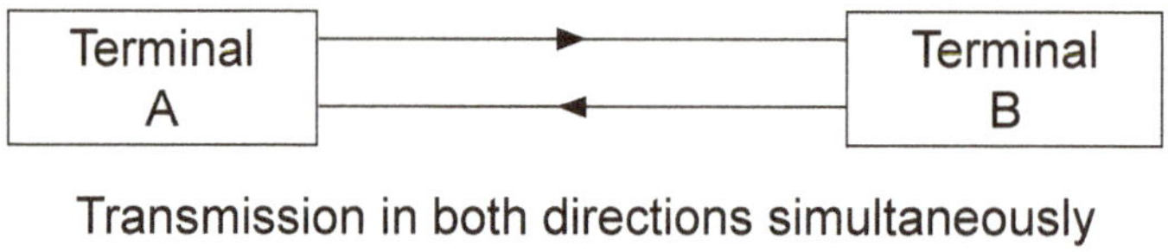

Diagram 5.19 Full Duplex Transmission

D. Transmission Techniques

Based on transmission techniques, a transmission can be classified into two categories: (a) Broadcasting (b) Switching

Under ***Broadcasting***, data is transmitted by one node and is received by many or all nodes.

The three switching techniques that are commonly used are..

In case of ***Switching***, the data is transmitted from the source to a specific destination.

The three switching techniques commonly used are (1) Circuit switching (2) Packet switching and (3) Message switching.

Circuit switching

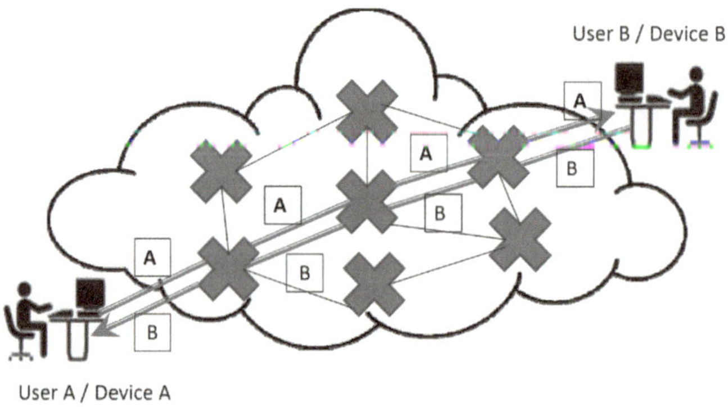

Diagram 5.20 Circuit Switching

In case of circuit switching, before communication occurs between the sending and receiving devices, a circuit is established if it is a connection-oriented transmission. Once the connection is established, the data transmission occurs over the established path even though alternative paths are available. The path is dedicated and exclusive when the communication happens and released when the communication terminates. An example of circuit switching is our home phone network.

Packet switching

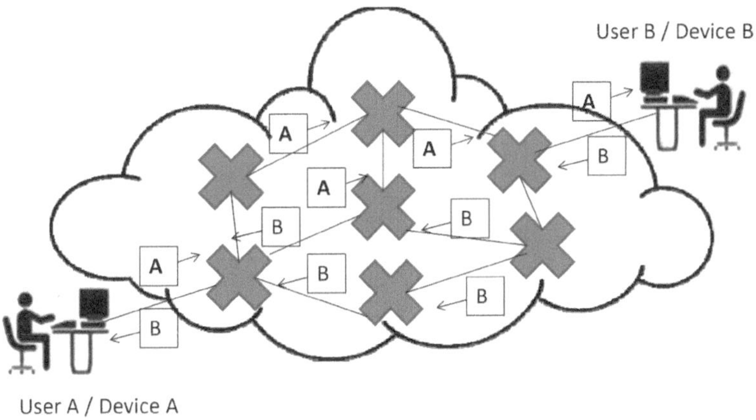

Diagram 5.21 Packet Switching

Packet switching is a connectionless data transmission, where there is no path setup before communication and the transmission is in full or in blocks. Each block of transmission may take a different path to reach the destination. Blocks will contain the destination address along with the message, which enables the block to reach the destination. This is comparable to the postal network.

Message switching

In case of message switching, the entire message that is being transmitted is stored in every node or switch through which the message is first routed and then forwarded. This is also called 'store and forward switching'.

Intermediary nodes or switches have the responsibility to pass on the message to another node in the network till the destination is reached.

Electronic mail transmission is of a message-switching mode.

5.4 Protocols and Internet Architecture

The main objectives of a efficient network architecture are to promote an open, simple and flexible telecommunication environment. These objectives are achieved through the standards set by quality organisations combined with the industry's best practices:

1. Standard Protocol
2. Standard hardware and software interfaces and communication
3. Standard multilayer interface between end-users and systems

Protocols

A protocol is a formal set of rules and procedures for communicating that includes the timing of message exchanges, the type of electrical connections used by devices, standard error-handling procedures and techniques, means of accessing communication channels, etc. The goal is to provide more standardisation and compatibility among the different communication protocols.

A. The OSI Model

In the year 1984, the International Standard Organization developed a standard for the Open System Interface (OSI) model for a secured data transmission. The data has to go through seven layers as detailed below:

Imagine **seven stages of product development** in a manufacturing line in the computer. At each layer, certain things happen to the data that prepare it for the next layer. The seven layers, which separate into **two sets**, are the application set and the transport set.

Application Set

Layer 7: Application

1. Using applications, the user invokes commands for execution by the systems and user interaction happens in this layer. The application could be a simple Word document creation or could be a complex analytical processing.
2. Application program or configuration contains all the information for establishing connection with intended partners.
3. The application also, in general, will have all error-handling mechanisms and data integrity checks.
4. The application takes care of all the protocols needed for working on the operating system and on the communication network.

Layer 6: Presentation

1. This layer handles the communication between the application and the layer downstream.
2. It translates the information readable on the application layer.
3. It handles activities such as translation, compression, encryption of data.

Layer 5: Session

1. The session layer is responsible for establishing, managing and terminating communications between the hosts.
2. It synchronises the dialogue between the application layers, between the hosts and manages data exchange.
3. It provides the service for an efficient data transfer.
4. It also handles exceptions-reporting of the session, presentation and application layers.
5. Based on the capability of the device, the session layer provides simplex, half-duplex and fully duplex communication between devices.

Transport Set

Layer 4: Transport

1. The transport layer essentially manages the flow control of data.
2. It defines segmentation, transportation and reassembling of data between hosts.
3. It establishes, maintains and terminates the circuits between hosts.
4. It may use either the Transmission Control Protocol (TCP) or User Datagram Protocol (UDP).

Layer 3: Network

1. The network protocol handles the method of sending data to the recipient devices.
2. This layer handles logical protocol, routing, addressing and path selection between two hosts either on the same network or on two different networks.
3. It moves packets (data) between one to the other end of the communication network.
4. This layer requires logical addresses such as an IP address.

Layer 2: Data

1. The data layer provides a definition of how the data has to be formatted for transmission.
2. It also defines the method of access to physical media and control.
3. Error detection, error correction, data recovery are the activities performed by the data layer.
4. It moves frame-to-frame between nodes.
5. It generally requires the MAC address of the devices.
6. It uses Ethernet protocol or Point-to-Point protocol.

Layer 1: Physical

1. The physical layer defines mechanical, electrical, procedural and functional specifications

for activating, maintaining and deactivating physical links between end-devices.

2. It ensures transmission of data bit-by-bit between devices.
3. It handles the translation of all upper layer data into binary code (0s and 1s).

The OSI Reference Model is mainly for a guideline. Actual protocol stacks often combine one or more of the OSI layers into a single layer.

How does Data move through the OSI Model?

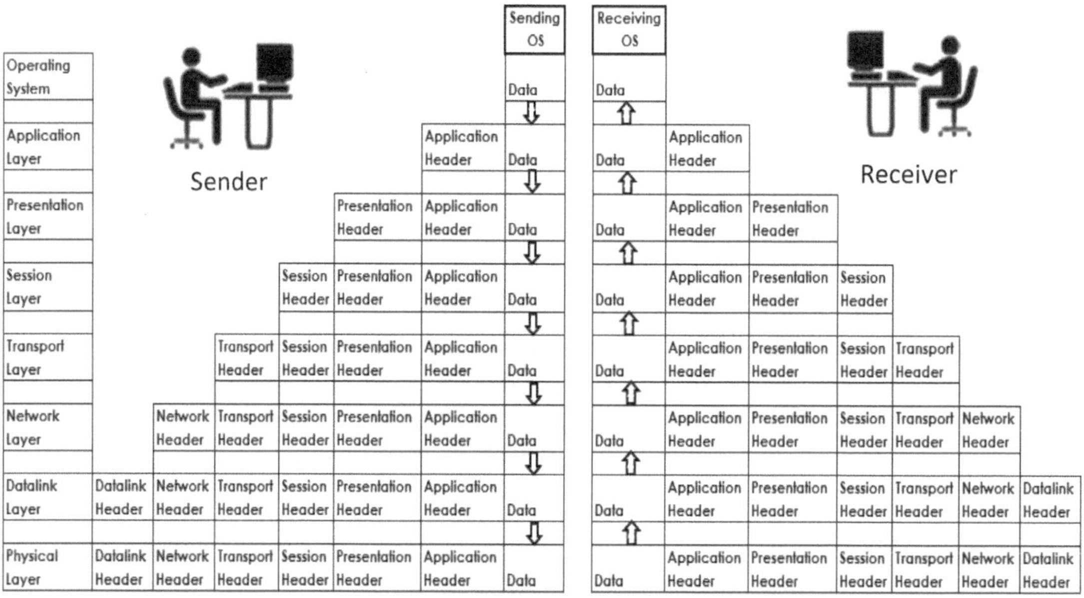

Diagram 5.22 Data moving through the OSI Model

B. Transmission Control Protocol/Internet Protocol (TCP/IP)

TCP/IP were the first networking protocols defined in this standard. TCP/IP provide end-to-end connectivity specifying how data should be packetised, addressed, transmitted, routed and received at the destination. This functionality is organised into four abstraction layers, which are used to sort all related protocols according to the scope of the networking involved.

From the lowest to the highest, the layers are

The Network Interface layer, containing communication technologies for a single network segment (link).

The Internet layer, connecting hosts across independent networks, thus establishing Internetworking.

The Transport layer, handling host-to-host communication.

The Application layer, which provides a process-to-process application data exchange.

TCP/IP and OSI Layers Compared and Matched

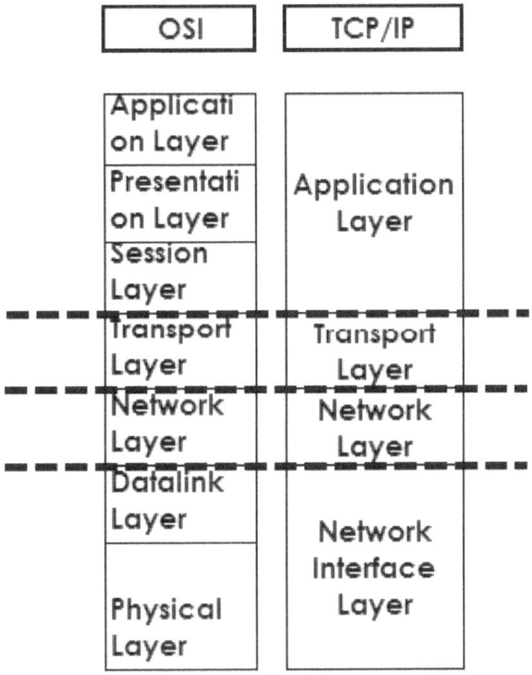

Diagram 5.23 TCP/IP Compared with OSI

Important Note: There is a school of thought, where the Network Interface Layer is split into two different layers: Network Interface Layer and Physical Layer. Certain educational institutions define these layers a little differently—5. Application Layer 5. Transport Layer 5. Internet Layer 2. Network Interface Layer and 1. Physical Layer.

In this case, Data link Layer/Data Layer of the OSI corresponds to the Network Interface Layer of the TCP/IP and the Physical Layers of both the models are matched. Students are advised to follow the respective institute material for examination purposes.

5.5 Network Risks, Controls and Security

As enterprises leverage more and more open Internet for conducting business, security risks are a major threat that need to be taken care of. Enterprises respond with ensuring adequate security for (1) safeguarding assets of the enterprise—assets includes both physical and information assets and (2) data integrity to ensure that in an automated data environment the data flow into the enterprise system is genuine and transmitted by the business partner.

In order to ensure security, the enterprises typically create a boundary system between main system and users. The boundary system provides user interface to access the main system resources. Security systems are implemented in the boundary systems. Boundary systems control the users' access and authorisation. Access is gaining entry into the IT environment for performing computing activities and authorisation is the ability to perform a certain set of activities. For example, user identification and password credentials are for gaining access to the IT environment. Upon gaining access, all users will not have the ability to perform computing for all the activities. The activities are allowed or restricted based on one's role in the enterprise. This is called role-based authorisation.

By security, we mean both (a) physical security of the enterprise premises and facility and (b) logical security, such as gaining access to the computing environment.

Threats and Vulnerability

Threats

Threats are classified into four categories: Structured threats, unstructured threats, internal threats and external threats.

Structured threat is something similar to organised crime. There are individuals, who are extensively knowledgeable and experienced about networking technologies, protocols, security, etc. who systematically penetrate and hack an enterprise's IT environment with or without malicious intentions to steal the enterprise data and information. Sometimes, industry competitors or state agencies hire these professionals for service.

Unorganised threats are by some technical enthusiasts who break and hack the security of an enterprise-computing environment. Generally, in these cases, the individual performs hacking just for thrill and not with a malicious intention for stealing data.

External threats are from individuals outside the enterprise; causing threat with or without malicious intention. These people will not have any user identification and other credentials provided by the enterprise. But, they still gain access to the enterprise-computing environment breaking the security.

Internal threats are from employees or ex-employees of the enterprise, who had been provided with user identification and other credentials. In this case, it is the misuse of one's position within the enterprise. Often, these cases are with malicious intentions.

Vulnerability

Intended or unintended flaws and weaknesses in setting up the computing environment are called vulnerability. A vulnerable IT environment is susceptible to threat. The following are the root causes of vulnerability:

Software bugs are a common vulnerability experienced by most of the enterprises' IT environments. Generally, these are unintended and quiet often, programmers fix it or find a way to work around this. Examples of software bugs are a failure in handling exceptional error-handling or memory overflow or failure to validate the input data, etc.

Timing windows is a method where an internal or external hacker gains access to a temporary file of an end user in the enterprise and after manipulating the content of the temporary file gains further access to the system.

Insecure default configuration is another method by which vulnerability is created. Instead of consciously configuring settings, the default set-up may expose the system's weaknesses, which are exploited by the hackers.

Untrustworthy information is another common method. Often users will receive mails from unsolicited sources with an attachment and when the user curiously opens the attachment, the attachment runs a certain script and the system set-up will change, providing access to the hackers who have sent the mail.

End-users are often the reason for causing vulnerability. For example, writing down the access credentials enables others to use the credential and gain access to the enterprise computing environment. Similarly, users may share credentials and eventually they get leaked to the outsiders.

Security Implementation

An enterprise's security administration is responsible for ensuring the running of the ongoing programs required for the enterprise's IT security. Generally, these programs are carried out through a periodic review of the security status and ongoing end-user education, evaluating technical advancement in security applications, etc. A security program involves the following steps:

Preparing the Project Plan for Enforcing Security: A security implementation project program will have goals, scope, tasks, milestones, etc. with the timeline of the program and resources needed for the program.

Assets Identification is the next step to decide and select assets such as personnel, computing assets, intellectual assets, facilities and machines to be secured.

Asset Valuation is the third step. Valuation of all the identified assets has to be performed. As insurance can be one of the methods of hedging certain risks, valuation can be done to compute the value of the sum insured.

Threat Identification is the fourth step. In this step, all probable threats are identified. The threat could be internal or external or structured or unstructured, intentional or accidental. Whatever their nature maybe—all threats are identified.

Threat Prioritisation: In this step, each identified threat is evaluated in the light of the probability of occurrence and the impact of threat—if it occurs—on the business. A combination of probability and impact analysis helps the enterprise to prioritise the potential threats to be responded with a counter-action.

Some schools of thoughts split the threat prioritisation steps into two: (a) Threat probability analysis and (b) Threat exposure analysis. But in effect, all the schools of thought aim at the same objective of planning on how to counter the potential threats.

Evaluation of Existing Control is the next step to verify whether the existing controls are sufficient enough to counter the potential threats. If not, the planning needs to be done for countering the un-hedged threats. This may call for the acquisition of new security assets and for the implementation of security measures.

Preparing the Program Report is the final step. In the report, the security administration department details all the analyses about the threats and their recommendations for review by the management and its approval for investment in new security measures.

Security Incident Management

In certain countries, there is a regulatory need for enterprises to immediately respond to a security incident and disclose the incident to all the affected parties and key stakeholders. For example, in the United States of America, the employee personal data and sensitive personal data are to be secured by the enterprise. In case of some security incident resulting in a compromise on this sensitive information leading to the exposure of personal sensitive information, the enterprise has to respond to such a situation with immediate corrective action. In addition, the enterprise has to report to all employees about the occurrence of the exposure of their sensitive information and personal sensitive information and what actions have to be taken by the management.

As a preemptive measure, every enterprise will conduct a security surveillance by scanning all files, system directories and the entire system environment. If any potential threat is suspected, the security officer is empowered to take the necessary proactive steps to prevent the occurrence of any security incident.

Network Security Protocol

When the data is transmitted on media, it is exposed to the open world. Intelligent and experienced IT professionals can intercept the message and interpret the content, unless the information in transit is secured. Therefore, there is a need for securing the information in transit.

Cryptography and encryption techniques help in securing the information in transit and only intended parties can read the information.

Cryptography is the practice of the study of techniques of secured information transmission.

Encryption: In cryptography, encryption is the process by which the information to be transmitted is encoded in such a way that the intruder or hacker cannot read or interpret the intruded information.

Terminologies

Plain Text

The information or message that has to be encrypted is called plain text.

Cipher Text

The plain text once encrypted is called cipher text. The cipher text has to be decoded to get the plain text again.

Decryption

Once the intended recipient receives the encrypted message, a process is to be followed to get the plain text from the encrypted message or information and that process is called decryption.

Encryption Model

An encrypted message or information cannot, normally, be read by the intruder. But, certain intruders can break the cipher.

Passive Intruder

A passive intruder is one who just listens to the intruded message or information.

Active Intruder

An active intruder is one, who just does not stop listening to the message, but also injects additional information into the message or modifies the message and sends it to the intended recipient.

Cryptanalysis

The art of breaking the cipher is called cryptanalysis.

Cryptology

Cryptography and cryptanalysis are collectively called cryptology.

In addition to the encrypted message or information in transit, there is a need for static messages and information to be encrypted in order to prevent the misuse of information by hackers. It is achieved in two ways: Hardware encryption and Software encryption.

Hardware encryption is implemented in several business enterprises to protect the information in the hardware. For example, a process called whole disk encryption is implemented using a software called PGP (Pretty Good Privacy).

Software encryption is generally done in conjunction with specific software applications. For example, a certain electronic mail software package offers encryption and decryption functionalities for mail security.

Network Security Protocols

SSH stands for Secure Shell. SSH is a program for logging on to another remote computer over the Internet and for performing certain activities. SSH provides a strong authentication and secure communication over the Internet. It protects the network from attacks such as spoofing.

SFTP stands for Secure Shell File Transfer Protocol. The security functionality is the same as that of SSH and this protocol facilitates file transfer.

HTTPS stands for Hypertext Transfer Protocol Secure. HTTPS encrypts the data flowing between server and client.

SSL stands for Secure Socket Layer. SSL is the protocol that provides a secure channel between two machines. SSL is extensively used for highly sensitive transactions such as credit-card payment process.

Network Security Techniques

As more and more of the automated data flows and self-service models such as vendor self-service, customer self-service, employee self-service models emerge, the usage of public Internet explosively increases for every business enterprise. This in turn poses a threat to assets, data and information. There is a strong need for protecting the enterprise information network from external threats and there are different tools and techniques available for this purpose.

Firewall

Firewall is a device that secures an enterprise network from the open Internet environment. Access controls are the common form of security control. Firewall can be called as a boundary subsystem. A user has to cross this boundary to access the enterprise's IT environment.

Message Authentication Code

Message authentication code is part of cryptography, where the code confirms the authenticity of the origin of the message. This is one of the reasons why the enterprises emphasis that official mails should not be used for personal communication. Usage of official mail id for personal purpose will land the enterprise in jeopardy.

Site Blocking

It is the software that prohibits users' access to certain sites by:

Intrusion Detection Systems (IDS)

IDS is a software that monitors network activities or system activities for malicious activities.

Network Intrusion Detection System (NIDS)

It is deployed on the network to monitor and analyse the network traffic for malicious activities' detection. This is also called as 'Packet sniffers'. NIDS is capable of intruding the packets and sending them to the specific host.

Host-Based Intrusion Detection System (HIDS)

It is deployed on the server to monitor, detect and respond to the user activities and external attacks on the host. HIDS primarily monitors the activities on the host.

Hybrid Intrusion Detection System

It is a combination of both network intrusion detection and host intrusion detection as a central intrusion detection system.

Network Node Intrusion Detection System (NNIDS)

In case of NIDS, the system detects the malicious packet intruders and sends them to the host. The NNIDS performs packet sniffer after the packet has reached the host. This is advantageous for hosts where HIDS is not deployed. NNIDS can detect intrusion only on the hosts it resides in.

5.6 Network Administration and Management

Network administration and management is a critical activity in enterprise computing. This involves monitoring network operations, administering the network, maintaining the network and service provisioning for the network.

Network Operations and Monitoring

It involves monitoring of traffic, spotting problems and troubleshooting issues in the network.

Network Administration

Administering the network involves having control over network assets and devices and periodic activities such as analysing the network traffic and reporting security issues to the management, etc.

Network Maintenance

It is the activity of a periodic review of network technology status, applying service packs, patches and upgrading the network to a higher version and migrating the network from legacy software to modern software, etc.

Network Service Provisioning

Adding users to the network, removing the users, providing authorisation to certain resources and removing authorisation, adding and removing nodes are typical activities of network service provisioning.

Network Management

Network management activities are popularly defined through an acronym called FCAPS. It stands for Fault, Configuration, Account, Performance and Security. These activities are a central part of network management.

Various varieties of activities performed by a network management team are: planning, controlling, allocating, deploying, coordinating and monitoring resources on the network, network planning, frequency allocation, predetermined traffic routing for load balancing, cryptographic distribution authorisation, managing configuration, managing faults, managing user accounts, managing security, managing bandwidth and route analysis.

5.7 Internet Revolution

There is no limit to writing about the Internet. While IT succeeded the Industrial Revolution, Internet accelerated benefits of information technology and with newer and newer benefits, models evolved from the Internet; it has become a revolution by itself.

The features of Internet and how businesses and e-Business models have leveraged that are detailed below:

Internet has enables the Overcoming of Geographical Barriers

Business Value	e-Business Model
This feature has helped businesses to connect operations in remote locations. Has enabled enterprises to evolve uniform business processes (say, vendor self-service or customers self-service) for all the customer and vendors all over the world.	Has enabled enterprises to reach out to customers in different parts of the world. Entrepreneurs were able to set up virtual business houses and mobilise logistical resources at the back-end.

Internet has enabled the Overcoming of Time Barriers

Business Value	e-Business Model
Irrespective of the time zones, business houses has been able to transact business and perform transactions in all parts of the world through the Internet	Online payment transactions and payment businesses are huge value for e-Business and have helped e-Business owners to cover the entire world; subject to the logistical tie-up one has.

Internet enabled the Overcoming of Cost Barriers

Business Value	e-Business Model
Internet has enabled communications through VoIP. This has helped enterprises to save significant costs, yet continue to avail benefits of international communication over the Internet.	Through VoIP technologies, several telephony service providers have offered services to the end-customer.

Internet enabled the Overcoming of Structural Barriers

Business Value	e-Business Model
Through business process reengineering, businesses have been able to set up 'business-to-business' processes and the business processes are cutting across the functional structure in an enterprise.	Business-to-customer is the model, which has helped businesses to remove barriers of the channels of distribution and convert their sales channels logistics and post-sales service purposes.

Internet and Other Networks

As we saw earlier in this section, a network in the context of IT is a collection of computers interconnected with one another. If such a collection is restricted to a certain entity, it is called Local Area Network (LAN). If multiple LANs of the same entity connected to each other spread across different geographical areas it is called as the Wide Area Network (WAN). So as, if a network is setup for a city or town, it is called Metropolitan Area Network (MAN).

Similar to a WAN, there can be two networks interconnected and transacting businesses of two different entities. It is known as **Extranet.** Typically, extranets are set up for an enterprise for transacting with vendors and customers.

Within a LAN, if the network is made available to the members of that entity for collaboration, knowledge-sharing and communicating, it is called **Intranet**. Internet is the backbone for the WAN with VPN, MAN and Extranet.

Internet Architecture

Internet is a World Wide Web. In short it is called WWW. For Internet, there is no central governing agency or any headquarters from where it is being controlled. Just by connecting to the Internet a computer can become part of it. Yet, the Internet has an amazingly naturally evolved architecture. Consider the diagram given below:

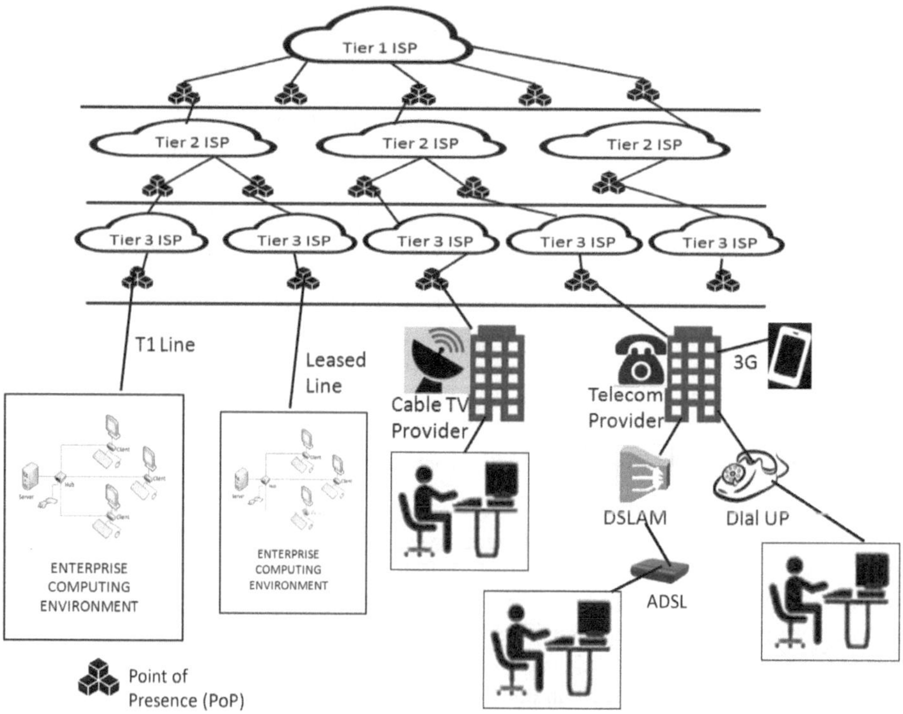

Diagram 5.24 Overview of Internet Architecture

(1) In the bottom-most layer of the above diagram, the users of Internet are shown. There are multiple ways in which a user can connect to the Internet:

1. **PSTN** is the acronym for **Public Switch Telephone Network**: a service provided by a telephone service provider. The access method could be
   a. **Dial-up** connection, where a user dials up a telephone number from a computer and gets connected to the Internet.
   b. **DSL (Digital Subscriber Line)** that has been connected to the computer through **MODEM**, which converts digital signals to analog for transmission through the phone line. DSL connections can be
      i. **SDSL (Symmetric Digital Subscriber Line)**, where the speeds and limits for uploading and downloading are identical.
      ii. **ADSL (Asymmetric Digital Subscriber Line)**, where the speeds and limits for uploading and downloading are different.
2. **CMTS (Cable Modem Termination System)** is another channel through which a user can access the Internet. CMTS is generally an Internet service provided by cable TV operators, leveraging their existing cable TV network infrastructure.
3. **T1 Lines** are generally a high-speed broadband Internet connection opted by business enterprises, where many numbers of Internet users will be consuming the Internet service.
4. Mobile phone/smartphone **3G service** is another way to access the Internet where the smartphone wirelessly connects to the nearest **POP (Point of Presence).**

(2) The second layer above the bottom-most layer is the Internet service provider to the end-customer. These enterprises can be a basic telecom service provider. At their premises a device called **DSLAM = Digital Subscriber Line Access Multiplexer** is installed to convert the analog Internet signal received through the phone line. Most of the **ISPs (Internet Service Providers)** who offer the Internet service to end-customers are Tier 3 ISP. They will have, in general, local network only.

(3) The third layer is, Tier 3 ISP which connects to the Tier 2 ISP. Tier 2 ISP will have Internet network in a given region in a country. The connection between the ISPs is happening through the **IXP (Internet Exchange Point).** This is called Peering. ISPs route the Internet packagers, based on the peering choice.

(4) At the topmost layer, Tier 2 ISP connects to Tier 1 ISP are players like AT&T, NTT Data, Tata Communication (including Teleglobe), etc. Tier 1 ISP has Internet infrastructure set up for the entire country.

Internet Applications

There are innumerable Internet applications and usages. Broadly, this can be classified into the following categories:

Information Acquisition

Getting information about a subject is easier over the Internet. Searches like Google, Yahoo, Bing help anyone to get from a basic to an advanced level of information on a subject.

Communication

Electronic Mail

Electronic Mail or email in short, is a way to communicate over the Internet.

Online Conversation

Internet applications like Skype of Microsoft, Face time of Apple help people to have online real-time conversation over the Internet.

Collaboration

Internet enables people with the same interest to collaborate. Virtual forums and communities are typical examples of collaboration.

Social and Professional Medium

Internet is becoming a social and professional medium through public services such as Facebook, LinkedIn and Twitter etc.

Business and Commercial

There are various models under which business transactions are being performed through the Internet such as B2B, B2C, etc.

Governance

Governments of various countries and states leverage Internet for governance. This is popularly known as e-Governance.

Business Uses of Internet

There are varieties of ways in which the Internet is leveraged by businesses. Some of the examples are as below:

1. **Strategic Alliances**: For example, a global e-Business enterprise can have a strategic alliance with a local partner in a country and start selling products in that country. The local partner takes care of the logistical aspects of the e-Business.
2. **Customer and Vendor Transactions:** Businesses leverage Internet for transmitting data between the business enterprises. The relationship between business enterprises could be that of customer and vendor. Besides, enterprises encourage customer and vendors to use

self-service portals.

3. **Product and Service Exchanges:** Through Internet, a seller and a buyer can be brought together by an e-Business portal. This process is called aggregating the market players.
4. **Marketing, Sales and Customer Service:** Internet is extensively leveraged by businesses for marketing products and selling and through web interface customer services are also extended.
5. **Web-Based Computer Applications**: A new concept called 'Software as a Service' has emerged where the software is available on the web for an authorised customer to use it for a subscription, as against the traditional model of licensing software.
6. **Enterprise Communication and Collaboration**: Internet is leveraged with the support of VoIP for voice and video conferencing by enterprises.

Intranet

Intranet is a locally restricted communication network created using Internet technologies.

Internet technologies such as TCP/IP, HTTP that are leveraged and used for creating resources with an enterprise are called Intranet.

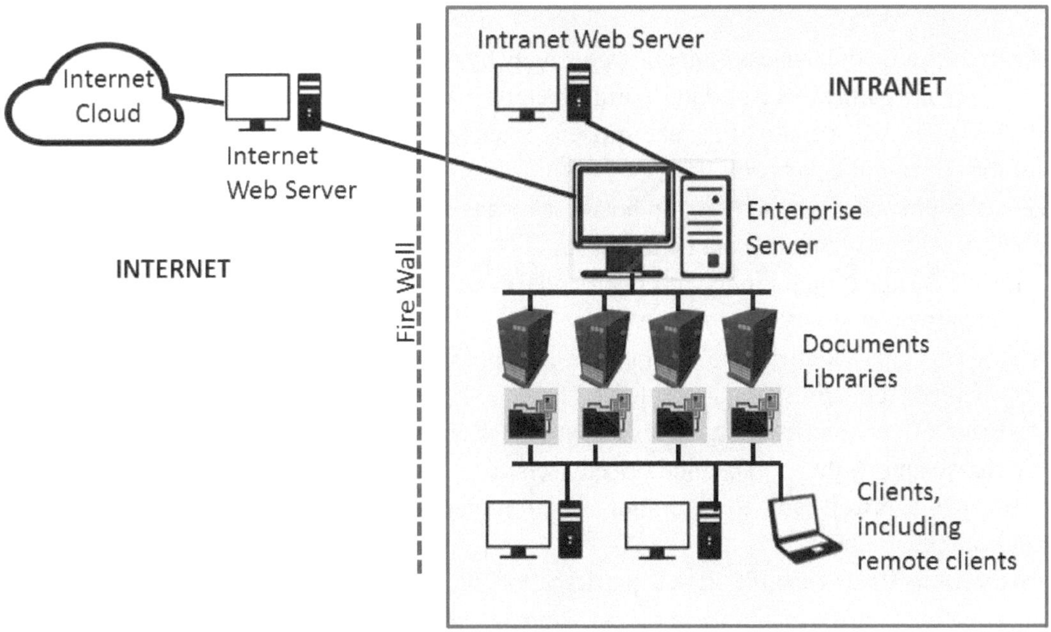

Diagram 5.25 Internet and Intranet

Intranet is extensively leveraged by enterprises for document management and knowledge management.

Also, a recent trend is to promote collaboration though social networking models coming up within the intranet.

Intranets are also leveraged for some of the HR applications based on the employee self-service model, where employees can submit tax declaration for payroll, submit expense reports for claiming official expenses, etc.

Managements of enterprises leverage intranet for employee communication, by publishing project success stories and recognition of employees, publishing enterprise newsletters, etc.

Extranet

Extranet can be defined as the interconnection two or more intranets of different enterprises. For example, an intranet of an enterprise is connected with the intranet of its customer or supplier; it allows the authorised user or customer or supplier to use the enterprise intranet.

The business value of extranet is enormous:

1. Normal commercial transactions between enterprises can be conducted through the extranet. For example, a vendor can submit an offer through the enterprise extranet of a vendor's customer. Similarly, invoices and other commercial documents can be submitted.
2. Through extranet, all the organisation employees can collaborate and develop products.
3. An end-customer can review the custom-made product under development.

Externalisation of business happens extensively by collaborating with business partners such as supply-chain partner, outsourcing partners etc. This offers a competitive advantage to the enterprise, as the value chain of the enterprise is extended to that of the supply-chain partner and that of the outsourcing partners.

The modern schools of thought on business management envisage the following distinctive advantages by deploying and using extranets:

1. Flexible business operations with open collaboration with customers and vendors, supply-chain partners and outsourcing providers.
2. Extranets helps enterprises to leverage the best possible benefits out of the Internet time that is procured from an ISP, such as a leased line.
3. Extranet offers security to the users of extranet, even though the data and information may be shared among the extranet authorised users from two different business entities.
4. Effective teamwork and collaboration are promoted among business partners with respect and confidentiality.
5. Extranet facilitates enable active participation of all members; thereby, every detail of information is made available to the decision maker. For example, if two teams of different organisations collaborate over an extranet and develop a product, minute details of the products are made available to everyone for developing the appropriate components of the product.

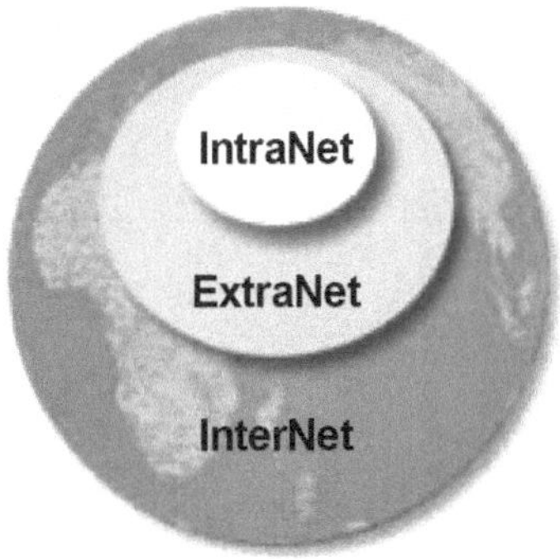

Diagram 5.44 Coverage and Comparison of Intranet, Extranet and Internet

5.8 Trends in Telecommunication

Telecommunication Trends: Industry, Technology and Application Trends

In the past couple of decades, the advancement in telecommunication trends has been pretty steep in all aspects.

Telecommunication Industry Trend

The telecom industry in various countries has witnessed deregulation and more players have entered into the business in each country. Such deregulation has also facilitated global players in expanding their global operations. Leveraging the technology convergence of communication, information and entertainment, telephony service providers also have also expanded their business offerings to Internet services and direct-to-home (DTH) satellite television channel service provisioning. Industries have leveraged World Wide Web for e-Commerce.

Telecommunication Technology Trend

Historically, telephony services had been based on analog and the shift towards digital has accelerated the technological advancement. Upon digitisation, ISDN (Integrated Services Digital Network) became possible, where one network is capable of performing, data, voice and video image transfer. Even during the digital era, the telephony service providers were having proprietary systems to provide telephony services but now the technology is open system. These systems have offered greater speed and interoperability between networks. On the network side, most of the providers are shifting from copper-wire based connectivity to microwave-based relay or fibre-optic connectivity.

An important technological trend witnessed by the telecom industry is the shift towards Internet-based telephony, technically called Voice over Internet Protocol (VoIP). The next important trend has been the wireless data transmission. NTT Data has experimented and succeeded in transferring data through wireless at a speed of 10GB per second. On the wired Internet service and data transfer side, invention of the fibre-optic cable accelerated the quality and speed of the data transfer.

Enterprises have started leveraging a high-speed Internet connection for conducting video conferencing and tele-presence. Smartphone features have compelled telephony service providers to advance their technology with respect to infrastructure. Telephone services providers have witnessed services over 2G, 3G, 4G LTE. Until recently, 4G LTE was used only for data transmission because of the technical limitations in smartphone handsets.

Telecommunication Application Trend

The past decade (2001–2010) witnessed a tremendous amount of business activities being performed over the Internet. Internet accelerated businesses and offered opportunities to entrepreneurs to evolve new business models such as e-Commerce. Almost all functions of a business have leveraged the Internet in conducting their business transactions.

Collaboration, speed, improved customer service, accelerated development of new products, cost savings and resource sharing are some of the benefits derived by the businesses leveraging Internet and communication services.

In the current decade (2011–2020), we are witnessing the mobile commerce as emerging and offering big benefits to the businesses. Mobile commerce (also known as m-Commerce) is possible primarily through the telecommunication network. Examples of m-Commerce are banking transactions, movie-ticket booking, cab-service booking, airline-ticket booking, etc.

Another big trend is mobile-based application development. This has become an industry by itself, with several thousands of players and billions of dollars worth of business.

5.9 Business Value of Telecommunication

Telecommunication: Business Value

Telecommunication technologies have offered enterprises to conduct their business over Internet under either the e-Commerce or the m-Commerce models. Both e-Commerce and m-Commerce have helped businesses to overcome several barriers:

- Geographical barrier
- Time barrier
- Cost barrier
- Structural barrier

Overcoming the above barriers, enterprises have derived benefits such as improved efficiency, consistency and reduced errors in conducting business transactions.

5.10 Telecommunication Networks

Telecommunication is the foundation of today's business communication. Telecommunication technologies are very complex and difficult to understand by laymen. However, there is a need for a basic understanding of telecommunication network technologies by everyone in the current business world.

Need for Telecommunication

Communication is an integral part of business activities. In order to successfully transact business, communication between employees and communication between business partners are vital. Through the telecommunication network the following activities are performed:

1. File Sharing: Sharing a common folder or files as part of collaboration between the people in the same team.
2. Resource Sharing: Sharing the printer resource.
3. Remote Access: Accessing an enterprise's IT environment by an authorised user connecting through the Internet and tunnelling the virtual private network with a secured user identification and password for performing official activities.
4. Shared Databases: Accessing a common database by multiple users.
5. Business Continuity: In a network, by implementing a concept called fault tolerance, data loss can be minimised or prevented. This is achieved through primary and secondary lines of back-up. Similarly, in case of a power failure, an uninterrupted power supply system ensures the functioning of the system. All these help in business continuity.
6. Secured access and data transfer through the Internet.

Telecommunication Network Model

Diagram 5.26 A View of Telecommunication Network

Over a telecommunication network, a sender sends a message over a medium or channel and the receiver receives the message.

In reality, the messages are passing through different devices and channels and processed by the telecommunication software, before being delivered to the receiver. All these are happening at a very high speed—say, a million transactions in a second.

The following are the five categories of components involved in any telephony messaging:

1. Terminal
2. Telecommunication processors
3. Telecommunication channels and media
4. Computers
5. Telecommunication software

1. Terminals

A terminal can be any device that initiates and sends a message or receives the message. These are end-points in the telecommunication architecture. The terminal device could be a computer or a video device or a smartphone or any office equipment such as a fax machine.

2. Telecommunication Processes

Telecommunication processors support the transfer of data between terminals. Network interface cards, MODEM, multiplexers and internetwork processors are called telecommunication processors.

Network Interface Card

Network Interface card is a part of the computer hardware that connects a computer to a local network. This has its memory and handles the outgoing and incoming packet transfer.

Diagram 5.27 Network Interface Card

MODEM

MODEM is the word coined from "Modulator" and "Demodulator" as the MODEM performs these two functions. A MODEM converts the digital packets received from a computer in a network into analog, which is called modulation and transfers the data to the recipient over the channel.

Diagram 5.28 MODEM

Multiplexers

A multiplexer merges the data transmission from several terminals into one and thus multiplexers enables a single communication channel to simultaneously carry data from many terminals. Similarly, data received for multiple terminals through a single line is separated by a multiplexer.

Diagram 5.29 Multiplexer

Internetwork Processors

As the name suggests, the devices under this category connect different networks. The following are called internetwork processors: switches, router, bridges, repeater and gateways.

Switches

A switch is a telecommunication processor that connects two telecommunication circuits.

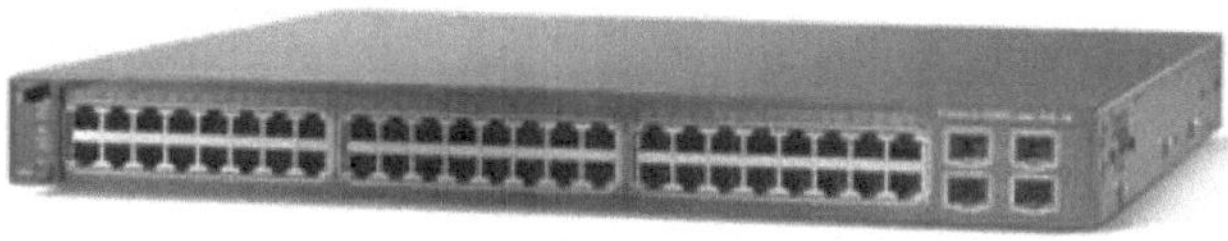

Diagram 5.30 Switch

Router

It is a device that also interconnects two telecommunication circuits and functions on a protocol.

Diagram 5.31 Router

Hub

A hub is another device, where multiple routers can be connected. Through it, multiple telecommunication circuits can be connected to a given circuit.

Diagram 5.32 Hub

Bridge

Bridge is the telecommunication internetwork processing device that connects multiple local area networks.

Diagram 5.33 Bridge

Repeater

Repeater is the device that boosts the signals before moving them to the next section of the network.

Diagram 5.34 Repeater

Gateway

A gateway is a communication processor that connects networks using different communication architectures. For example, in the United States, a telephony service provider may offer basic telephony services, high-speed Internet services and DTH satellite television services. A home that consumes all these services can have a gateway as a single telephony processor for all the three services as a gateway has architecture for handling all the three communication protocols.

Diagram 5.35 Gateway

5. Telecommunication Media/Channel

Telecommunication channel/media connects the message sender and the message receiver. The data is transmitted over the channel, which is a part of the telecommunication network.

The channel or media can broadly be classified into two categories: ***1.Guided media or bound media or 2.Unguided media or unbound media***

Guided Media/Bound Media

Guided or Bound Media is a type of transmission of data through a physically laid cable. Examples of guided media are twisted-pair cable, coaxial cable and fibre-optic cable.

Twisted-Pair Cable

Twisted-pair cable is an ordinary telephone line wire made of copper twisted into a pair. Though this is a widely used guided media for both data and voice at home and in office, there are disadvantages. This media is susceptible to electrical interference; hence, there will be noise. Over long distances, the data and voice are garbled; hence, a number of repeaters will be required.

Diagram 5.36 Twisted Pair Wire

Coaxial Cable

This is a media cable that is made of either copper or aluminum and wrapped with spacers that minimise the interference. Coaxial cables are very popular for cable television connectivity and cable television channels are one of the carriers of Internet services. This is also widely used at home and in offices. This medium is more expensive than twisted-pair wires.

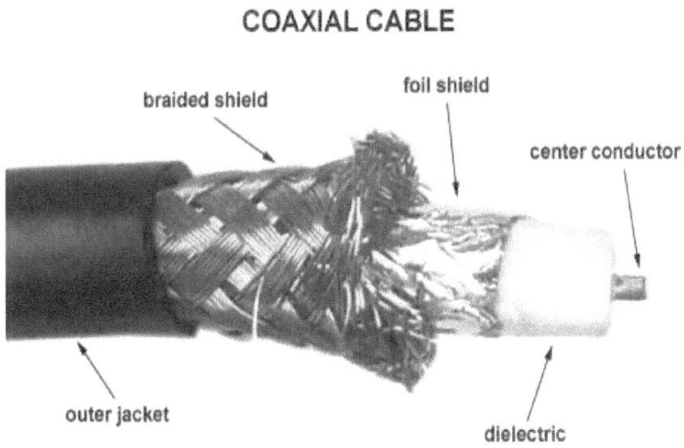

Diagram 5.37 Coaxial Cable

Fibre-Optic Cable

Fibre-optic cable is a media that consists of one or more hair-thin filaments of glass fibre wrapped and protected in a jacket. Signals are converted into light forms by laser bursts fired into this media. Fibre-optic cables are capable of carrying both digital and analog signals. The speed of transmission and the capacity to carry signals are much higher than the coaxial and twisted-pair media. This media is much speedier than the microwave satellite system. High investment needs are the only disadvantage.

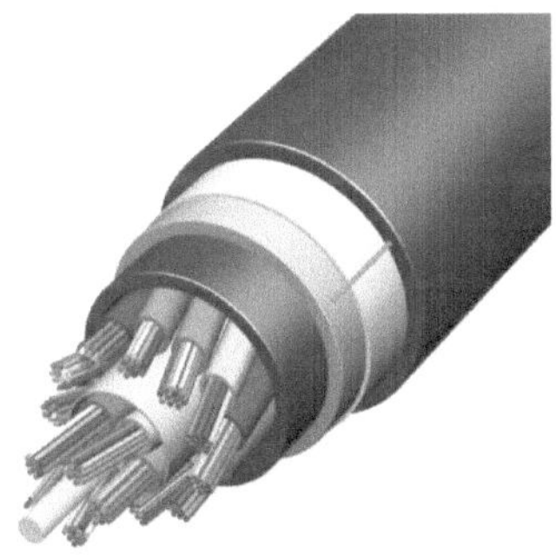

Diagram 5.38 Fibre-Optic Cable

Unguided Media/Unbound Media

Unguided media are those media, which are not wired media. In other words, unbound media are wireless telecommunication channels. The popular unguided media in telecommunication channels are terrestrial microwave, radio waves, microwaves, infrared waves, communication satellite.

Terrestrial Microwave

Terrestrial microwave is a medium used for transmitting data from an earth-bound microwave station to another destination where the line of sight is clear and there is no interruption between the earth station and the destination. The maximum distance that a terrestrial microwave can cover is 30 miles. The natural atmosphere is used for transmitting the microwaves. The biggest disadvantage of terrestrial microwave is it that it cannot be transmitted along the curvature of the earth; it has to be straight with the destination being in the line of sight without interruption.

Diagram 5.39 Terrestrial Microwave

Radio Wave

Radio waves are another unbound media through which the data is transmitted through invisible electromagnetic radiation. This is one of the widely used unbound media. Technically, the wavelengths of radio waves vary from 1 mm to 1,00,000 km Radio waves are widely used in a wireless local area network.

Diagram 5.40 Radio Wave

Microwave

Microwaves are electromagnetic waves ranging from 1 mm to 1 m with the frequency 0.3GHz to 300 GHz. This is widely applied in radar communication, radio astronomy and navigation communication.

Diagram 5.41 Microwave Network

Infrared Waves

Infrared waves are applied in industrial, scientific and medical applications. Infrared is also applied in remote-control devices used in controlling television operations and widely applied in guided-missile applications.

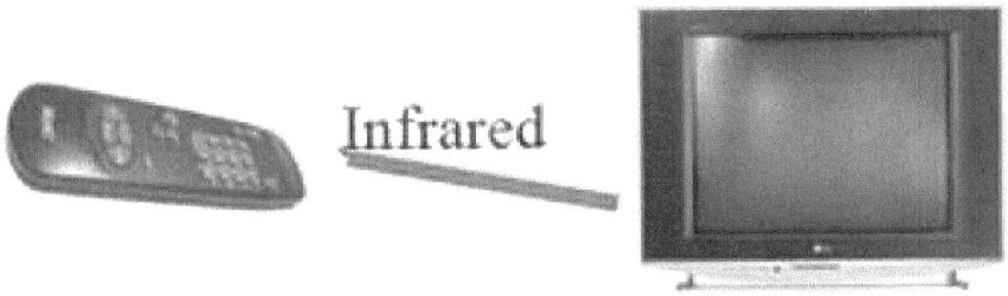

Diagram 5.42 Infrared Waves

Communication Satellite

Like terrestrial microwaves, communication satellites also use the atmosphere to send signals to earth stations. These are prominently used for high-volume long-distance data transmission. A satellite is a solar-powered device in the earth orbit that receives, amplifies and retransmits data to the earth stations. Meteorological activities leverage satellite communications extensively.

Diagram 5.43 Satellite Communication

5. Computers

Computers are an integral part of the telecommunication network. The types of computers used in telecommunication network are

- Mainframe computers are used as host systems as these systems need to handle a high volume of data.
- Microcomputers are used as network servers
- Minicomputers are used for front-end activities

6. Telecommunication Control Software

There are various types of telecommunication control softwares that are used, based on different types of computers used at different layers of the telecommunication network.

- Telecommunication network monitors generally operate on the mainframe system, which runs with proprietary operating systems.
- Microcomputers that are used for network management typically use network operating systems that consist of network-operating components and communication packages.

Telecommunication control software enables network management that involves the following activities:

Traffic Management

This functionality manages the network traffic and performs resource optimisation to avoid network congestion.

Security

This feature enables authentication, encryption, decryption, data security, auditing and enforcing the security policies.

Network Monitoring

This functionality enables the network administrators to monitor the network and troubleshoot if any issue is reported and also take a proactive approach to preempt a problem, if any of the network parameters are adverse.

Capacity Planning

Provides rich data and growth pattern to plan the future network capacity.

5.11 Electronic Commerce

As more and more applications of Internet on business happened, a new business model evolved called electronic commerce. Initially, the business models were only conceptual and as more and more entrepreneurs experimented and achieved success, more players entered. Today, the e-Commerce is a key force to reckon with.

It is the process of conducting business transactions electronically. Using this technology, the transaction processes were enhanced beyond geographical boundaries, beyond time, at an affordable cost, cutting across organisations and structures.

E-Commerce is an automated business process between business-to-business and business-to-customer. Based on the past experience, most of the schools of thought still believe that a greater integration of e-Commerce with mainstream business is yet to happen. In a nutshell, a business enterprise cannot afford not having an e-Commerce footprint as it is more than a necessity to compete in a competitive market.

Benefits of e-Commerce

1. Benefits to customers with a better price with better quality due to increased competition.
2. Reduction in error and overhead as data processing is automated at every stage.
3. Reduction in overhead cost of all entities as most of the processes are automated, as against manual submission in conventional business.
4. Speed in business and transaction processing.
5. Market expansion.
6. Market penetration.
7. Improved quality of products from the customer's perspective, as customised products can be produced and delivered, collecting self-configured product specifications from the customer.
8. Reduction in inventories; practicing just-in-time inventory management is possible.
9. Reduced paper work as most of the documents are electronically stored.
10. E-Commerce helps in publicity.

With the above benefits, there are also inherent risks associated with e-Commerce.

E-Commerce Risks

1. **Anonymity of e-Commerce Business Owner:** It is easier said than done to do business electronically, as trust can be earned only through the physical brick-and-mortar model. Further, with lots of failures of e-Commerce businesses in delivering promises, the trust level goes down with certain customers—if not all.

2. **Non-Acceptance of Business:** A business transaction may not be accepted by an e-Commerce provider.
3. **Transaction Authenticity:** At times, the transaction document authenticity is questionable in e-Commerce. There are players in the e-commerce business who have failed to fulfill their promise in spite of the provision of the so-called authentic order confirmation.
4. **Risk Posed by Intruders:** Personal and sensitive information such as credit card details may be intruded into and hijacked.
5. **System Dependency:** E-commerce is completely dependent on technology, systems and machines and when these fail, the service is unavailable to the customer.
6. **Legal Validity:** E-commerce transaction documents may not be admissible in all courts of law.
7. **Problem of Piracy:** Intellectual properties sold through e-Commerce may easily be pirated. For example, a copy of a PDF form of a book sold over e-Commerce can be distributed to multiple readers.

Types of e-Commerce

1. Business-to-Business (B2B)

This refers to e-Commerce transactions between two or more businesses. In this process, the businesses exchange transactions electronically. The transactions could be circulating a request for a proposal, submitting the proposal, placing a purchase order, providing commercial documents such as packing slips, delivery notes, shipping documents, transmission of receivable transaction to customers, electronic payment through bank, etc. The criteria for this case are that the entities involved are business entities, hence it is called B2B.

2. Business-to-Customer (B2C)

In the case of B2C, the business is conducting business transaction with customers. From displaying the product on the web catalogue, the entire transaction cycle is done electronically. Amazon, Flipkart are examples of B2C e-Commerce type.

3. Customer-to-Business (C2B)

In case of customer-to-business, a customer offers one's product or service to the business through an e-Commerce portal. For example, collecting reviews about product experience by customers and making those available to the business is a model which is prevalent throughout the Internet. For example, Priceline.com

4. Customer-to-Customer (C2C)

C2C is a kind of exchange where a customer offering a product is posting the product with an expected price and a needy customer may choose to buy the product. For example OLX.com is a C2C example.

5. Business-to-Government (B2G)

B2G is about government procurement and the service tendering process conducted through the Internet. Besides, e-Governance can also be part of B2G.

6. Business-to-Employees (B2E)

This is a business offering its services to employees as part of the employment terms, such as processing an employee request for vacation, expense claims etc.

Critical Success Factors for e-Commerce Implementation

Processes, Policies and Procedures

The enterprise has to completely map all processes and how to implement them. Policies relating to each of the business transactions and exception handling have to be finalised and a clear procedure has to be drafted for all the situational transactions.

Analysis

Analysing the cost-benefit analysis along with the risk analysis and the value proposed to be delivered from e-Commerce has to be clearly stated and articulated.

Security Consideration

All security considerations have to be taken into account and planned.

Evaluating Internal Controls

As part of the process mapping, internal controls have to be established between processes to ensure that nothing falls through crack.

Integrating e-Commerce with Back-End Logistics

Delivery capability of e-Commerce is most critical to earn the trust of the customers. This has to be planned as a part of the back-end logistics plan.

Training

It is imperative that every employee in the enterprise is trained on the e-Commerce business model.

Periodic Post-Implementation Review

After implementation of e-Commerce, there is a strong need for a periodical review for improving the process and tightening controls are needed.

5.12 Mobile Commerce

Mobile commerce is an extension of e-Commerce. As a result of the convergence of information technologies and communication technologies in a single device—a smartphone—most of the transactions which were hitherto conducted with the help of a computer, can be now conducted over mobile devices. This is known as Mobile Commerce.

Mobile commerce includes transactions conducted through laptops, tablet PCs, PDAs besides smartphones. Mobile Commerce is also called as m-Commerce.

To enable a mobile phone transaction, a protocol called Wireless Application Protocol (WAP) has to be set up.

The types of m-Commerce businesses are banking and financial service transactions, telecommunication business transactions, retail, information services, etc.

With the advent of the tablet PCs and with the smartphones becoming common among most of the users, m-Commerce applications have begun to explode.

5.13 Electronic Funds Transfer

Electronic Funds Transfer (or simply EFT) is a banking transaction process through which a business or an individual account holder in a bank receives deposits or effects payments electronically, instead of physical issue of cheques or withdrawal slips or deposit slips.

EFT transactions are much faster than the conventional payment processes.

The prerequisite for EFT is authorisation by the account holder for electronic payment.

Examples of EFT:

Automated Teller Machine (ATM)

ATM is a machine that enables a business transaction in a bank through the insertion of an ATM card (similar to a debit card) issued by the bank along with a Personal Identification Number (PIN) which to be set and provided by the account holder for every transaction. Once the user identification is validated with the ATM card and the PIN, the ATM machine executes the transaction and in the case of it being a withdrawal transaction from the bank, disburses cash.

Point of Sale (PoS) Retail Transaction

PoS retail transaction payments are something similar to ATM transactions, but in this case both the debit card and the credit card can be used for transaction. Once the transaction is validated, the financial institution makes the payment to the merchant.

Pre-Authorised Transfer

This is an electronic authorisation provided by the account holder to the bank for the transfer of money from one account to another (including a third-party account). For example, an account holder may authorise the bank to transfer money from a checking (or savings) account to a credit-card account.

Telephone Transfer

This is also called as phone banking, where the account holder authorises the bank to effect a transaction over phone, with valid user-identification credentials.

Section 6
Business System

6.1 Background

Business Information System (BIS) is comparable to the human nervous system in a business. BIS is not only critical to a business but also a need for its survival. While the BIS was becoming more and more powerful on its own, information and Internet technologies have added an accelerated power to it.

With the combined power of BIS, information and Internet technologies, entrepreneurs have evolved newer business models and thus the emergence of e-Commerce.

6.2 IT: The Key Business Enabler and Driver

Over the past couple of decades, the ways of conducting business have dramatically changed. Some of these dramatic changes are

1. Businesses have started embracing Internet technologies and the emergent e-Commerce.
2. Information Technology and Internet-enabled globalisation of enterprise operations can be seen from two perspectives:
   a. Reaching out to global customers.
   b. Reaching out for global resources for optimal costs and operations.
3. There is a shift from supplier-centric business operations to customer-centric business operations.

When such dramatic changes are happening due to various environmental dynamics and changes, every decision maker in a business needs the support of business data and information for appropriate decision-making. In other words, a decision maker needs to take a well-informed decision.

Business information systems enable decision-making, provide all business data and information and enable a decision maker to take a well-informed decision.

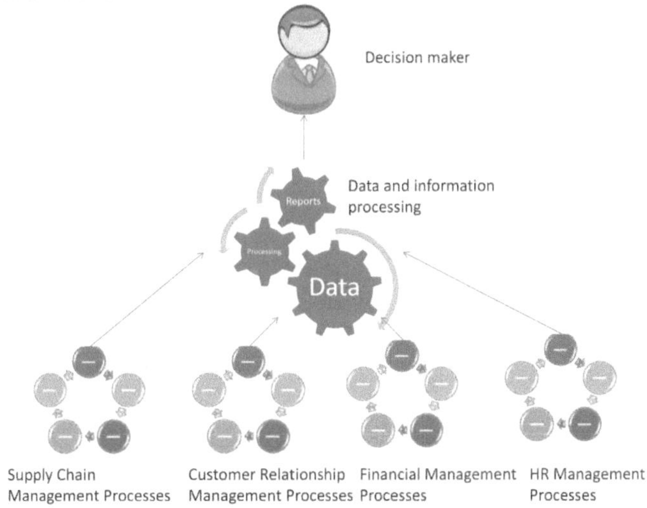

Diagram 6.1 Interconnected Business Process Data-Data and Information Processing-Decision-Making

Information System Interface: People, Process and Technology

Often, the terms information system and Information Technology are interchangeably used. But, it is incorrect to use these two as substitutes for each other; It is because "Information Technology" is not "Information System".

Information system is much larger than IT. As a matter of fact, IT is a small subset of the information system.

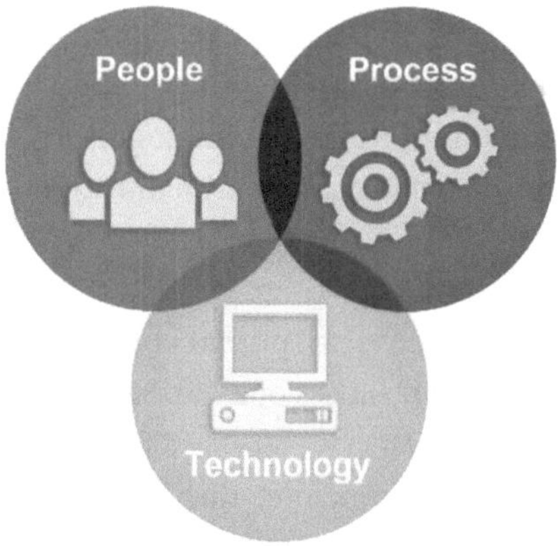

Diagram 6.2 Elements of Information System

Elements of an Information System

People, process and technologies are the elements of an information system. In a business enterprise, data is generated when the processes are executed for conducting transactions or are executed by people or triggered for execution by people. The generated data is then processed for information reports. For processing data and generating reports, information technology is used. Further, people use these reports and consume information for decision-making.

There is a school of thought where it is assessed that of the three elements above, 50% is 'people' as they trigger transaction and consume information; 'processes' are 40% as they generate data and 'technology' is only 10%.

People, process and technology in data generation and information processing are collectively called as information system.

6.3 Information System

Information, System and Information system

Information

An enterprise typically will have processes such as 'order to cash', 'procure to pay', etc. and if any manufacturing activities are involved then 'plan to produce'; for HR activities 'hire to retire' and for accounting 'record to report'.

When these processes are executed to conduct business transactions, an enormous amount of data gets generated in the functions of sales, distribution, procurement, inventory management, manufacturing, human resources and accounting.

For making meaningful information out of these massive data, we need a technology that processes the data and provides meaningful information.

People in operations are triggering transactions; people in the IT function work on Information Technology to transform data into information and people at the executive and management levels consume the information for decision-making.

System

A system is a collective and cooperative function of the different elements in the system, working towards a common goal, receiving inputs and processing inputs to produce outputs.

Information System

Information system is a combination of people, process and technology cooperatively working together, consuming the inputs (data) generated in the business and processing them for output (Information).

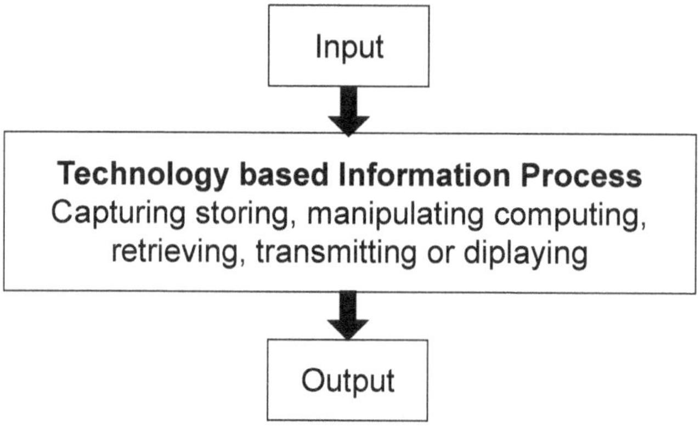

Diagram 6.3 Information System

Components of the Information system

The very purpose of information system is to convert input data into meaningful information output through data processing techniques.

People, data, software and hardware are components of an information system.

The definition of 'people' includes the end-users of an information systems and IT professionals who enable setting up the system for data processing. Operational-level end-users typically generate data running the business transactions or simply feed inputs to the information system and executive management uses the outputs for decision-making.

Process or data processing is used for conversion of inputs into meaningful information or output.

The information process consists of input of data, data processing and output of information, storage of data and the information and control process.

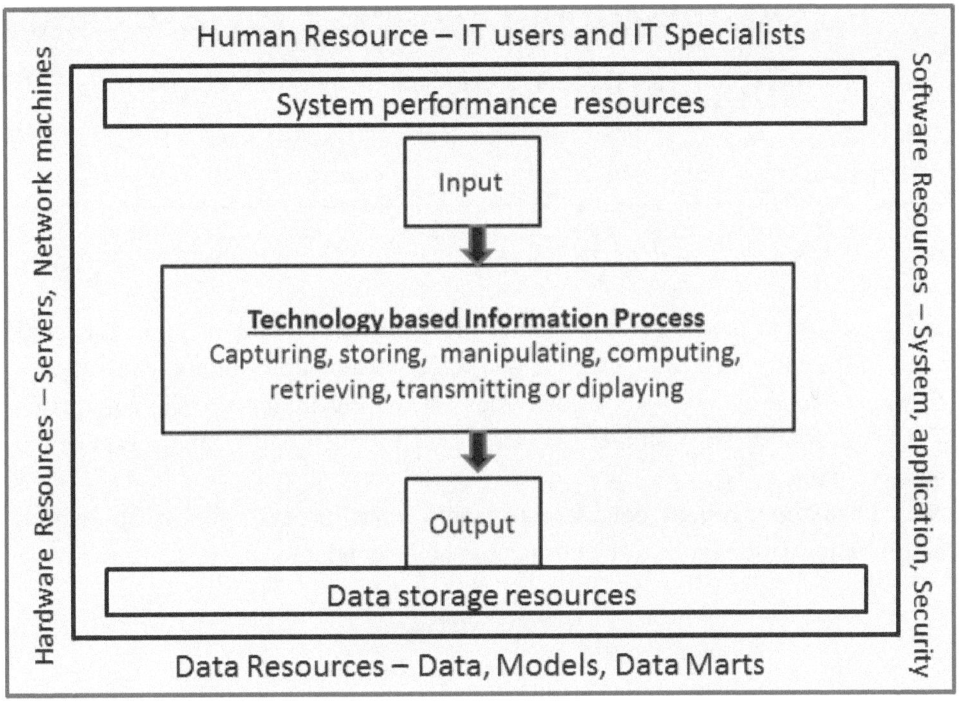

Diagram 6.4 Components of Information System

Business Information System

A business information system is the integration of information technology, people and business. The Business Information System (BIS) is enabling people to conduct business transactions efficiently and effectively and in capturing the generated data and processing it for generating meaningful information for timely, accurate and well-informed decision-making.

6.4 Organisation, Business Processes and Information System

An ***enterprise's*** goal is to serve a customer profitably. In order to achieve this goal, different functions of the enterprise come together and work towards the common goal.

Business processes are a set of activities carried out to achieve the enterprise's strategic goal i.e. to ensure that the customers are served well to their entire satisfaction and to make the business profitable.

An ***Information system*** consists of people, data, process and technology to produce meaningful information on how the business is being conducted.

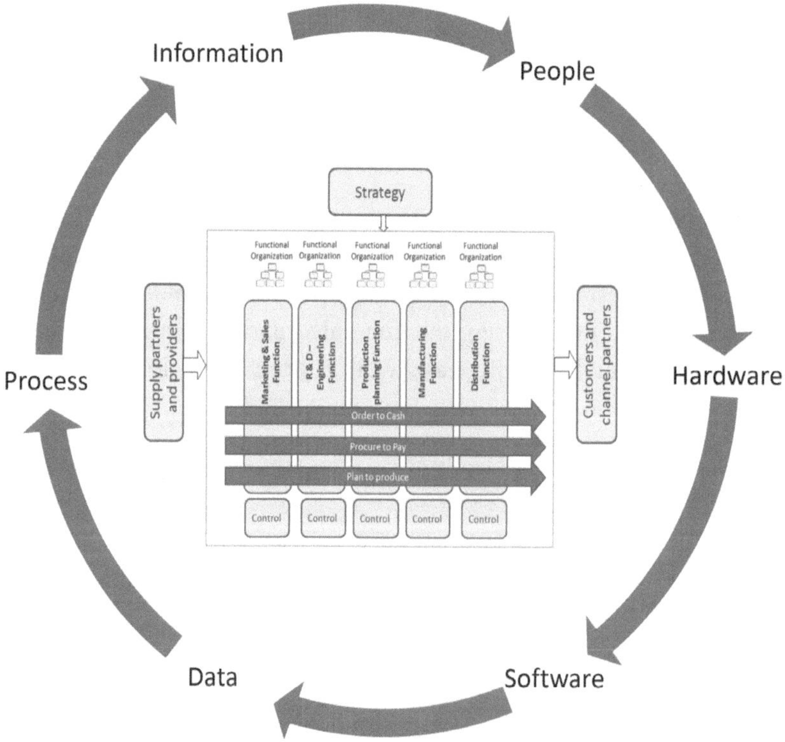

Diagram 6.5 Organisation, Business Processes and Information System

In the past, application of IT was optional for a business. In the current era of information and Internet technology revolution, application of information technology and setting up an information system is no longer an option, but a basic competitive need to survive in business.

Every resource in a business demands a liberal IT support for competing in the market, for gaining competitive advantage, better decision-making, speedier service and product delivery, faster exchange of information and for reducing information redundancy and inaccuracy.

6.5 The Information System's Role in a Business

It is a universally accepted fact that IT is an accelerator of the business strategy.

IT helps a business in

1. Managing the business processes better
2. Enabling competitiveness
3. Enabling information-based well-informed decision-making
4. An optimised utilisation of resources of the enterprise
5. Technologies like the Internet enable business growth through
    a. Market penetration
    b. Customer intimacy
    c. Collaborative supplier partnership
    d. Collaborative product development
    e. Operational excellence
6. IT helps an enterprise to achieve cost leadership and product leadership.
7. The operational level is benefited by business automation through a transaction processing system such as the ERP and by an office automation system for document processing.
8. The senior management is supported by the decision support system.
9. Senior management gets to effectively control business through the management information system.
10. Executives are able to have online executive systems such as balanced scorecard.

6.6 Types of Information System

An information system can support all levels of the hierarchy of an enterprise right from operations at the bottom to strategy-making at the top.

At the **operational level**, the system that is implemented to handle the day-to-day operations is called the *Transaction Processing System.* For example, ERP is an online transaction processing system.

The **support staff** needs to create the enterprise communications and presentations and prepare ad hoc reports. For this purpose, *Office automation systems* such as MS Office and MS Outlook for email are deployed in an enterprise.

In some enterprises, this support staff is classified as knowledge workers. *Design engineering* functionaries in an engineering enterprise can be considered as knowledge workers. For these types of functional needs, varieties of functional softwares exist. For example CADD—Computer Aided Design and Drafting—is a software for the design engineering staff. Similarly, in a chemical business enterprise, *LIMS (Laboratory Information Management System)* is a software that may be a critical need for knowledge workers.

At the **tactical management level**, the managers need periodic information about various parameters of operations and to take action based on the trends; for this *management information systems* are deployed.

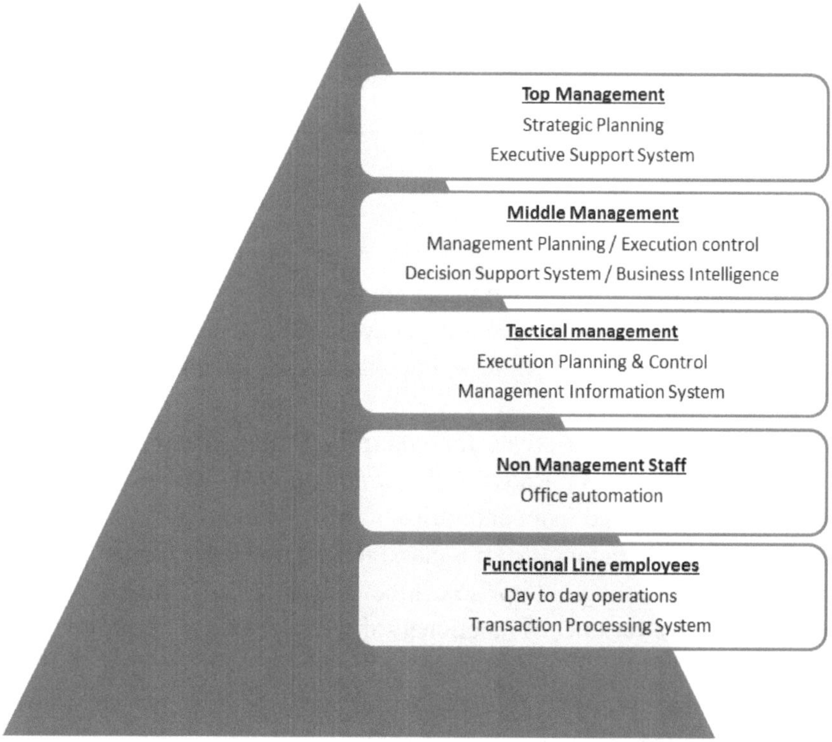

Diagram 6.6 Information System at Different Levels of an Enterprise

At the **middle management level**, senior managers need systems that aid their decisions. These are called *Decision Support Systems*. Decision Support Systems of the olden days have been replaced by the *Business Intelligence* systems in the current era.

At the **strategic management level,** the needs are strategic project monitoring, balanced score card and similar information solutions. The system utilised for fulfilling these needs is called the *Executive Support System* and it is also called as *Executive Information System.* In the current era, the Executive Information Systems are being fast replaced by the *Strategic Enterprise Management* systems, which enable the executives to monitor business performance online in real time.

Transaction Processing System

A transaction processing system collects, stores, enables modifications and the retrieval of day-to-day transaction data on the fly as the transactions happen. Transactions are happening across the processes for different functions. For example, sales-order processing, material issue for shipping, shipping transaction, invoice preparation and dispatch, receipt of payment from customers are processes in the 'order to cash' function. The data is collected as transaction happens on the fly.

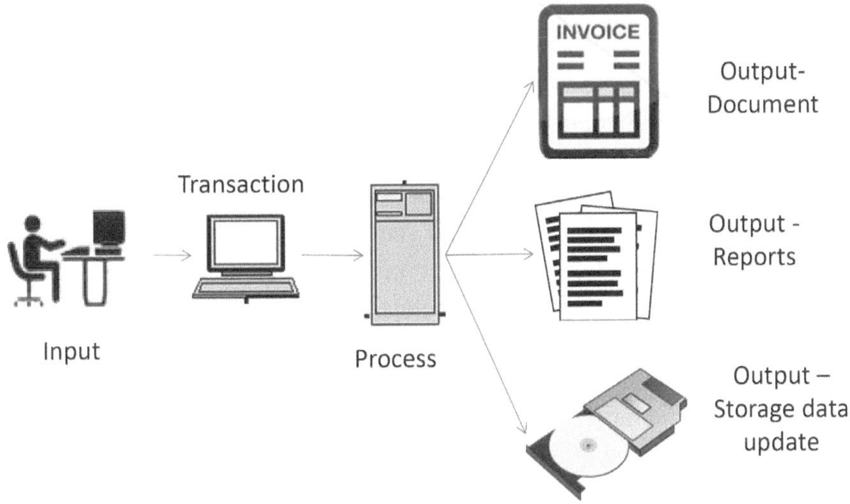

Diagram 6.7 Transaction Processing System

Characteristics or Attributes of a Transaction Processing System

The following are the characteristics or attributes of a transaction processing system:

Access Control and Authorisation: The access control to a transaction processing system is generally restricted. Generally, access is provided to a member of an organisation based on the role and the need. Further, the user identification and password have to be authenticated for an authorised user to gain access. As part of the audit trial, the systems keep records of the user who triggered the transaction.

Equivalence: Transaction data, as user inputs, is collected in a transaction processing system in a standardised format. This is done irrespective of the source of information. For example, if a customer order is to be created by the sales personnel using a terminal from the office or a customer creates one's own order through the self-service system over the Internet, the format (data collected) is one and the same.

High Volume and Rapid Processing: Transaction processing systems are capable of processing high volume transactions and at a rapid speed. This is because in a transaction processing system, granular-level data is captured and processed for output.

Trustworthiness: When a user inputs data into a transaction processing system, the data fed undergoes several validations before being accepted by the system for processing. Hence, the data is reliable.

Transaction Processing Qualifiers

For a transaction processing system to process a transaction, it has to go through certain conditions. These conditions are called ACID—an acronym for Atomicity, Consistency, Isolation and Durability.

Atomicity: As mentioned earlier, in a transaction processing system the data is gathered at a granular level. Therefore, the transaction is either considered in full in all respects or it is rejected by a system. For example, when you record an accounting transaction, the system will consider both debit and credit; only if there is no error. If there is an error in credit, even though all the data relating to debit is in order, the transaction will be rejected. This attribute is called Atomicity.

Consistency: A transaction enforces consistency in the system's state by ensuring that at the end of any transaction the system is in a valid state. If the transaction completes successfully, then all changes to the system would have been properly made and the system will be in a valid state. If any error occurs in a transaction, then any changes already made will be automatically rolled back. This will return the system to the state it was in before the transaction was started. Since the system was in a consistent state when the transaction was started, it will once again be in a consistent state. For example, in a bank account transfer system, the system is consistent if the total of all accounts is constant. If an error occurs and the money is removed from account A and not added to account B, then the total in all accounts would be changed. The system would no longer be consistent. By rolling back the removal from account A, the total will again be what it should be and the system back in a consistent state.

Isolation: When a transaction runs in isolation, it appears to be the only action that the system is carrying out at one time. If there are two transactions that are both performing the same function and are running at the same time, transaction isolation will ensure that each transaction thinks it has the exclusive use of the system. This is important because as the transaction is being executed, the state of the system may not be consistent. The attribute of 'Isolation' ensures that the system remains consistent even after the transaction ends, but during an individual transaction, this may not be the case. If a transaction was not running in isolation, it could access data from the system that may not be consistent. Transaction isolation prevents this from happening. Imagine a global ERP system that is processing sales-order transactions globally. At any point it time, there will be several transactions going on simultaneously and there will be no consistency when one transaction is happening. But at the end of the day, an application system ensures that all transactions in all respects will be updated in the database sequentially in the order of transaction.

Durability: A transaction is durable in that once it has been successfully completed, all the changes it made to the system will remain permanently. There are safeguards that will prevent the loss of information, even in the case of system failure. By logging the steps that the transaction performs, the state of the system can be recreated even if the hardware itself has failed. The concept of Durability allows the developer to know that a completed transaction is a permanent part of the system, regardless of what happens to the system later on.

Office Automation System

The office automation system helps in the automation of ***communication, office resource management, document creation and document management.***

Communication Management

Implementation of a software that helps in managing the automated mail communications and instant messaging helps in the management of automated communication. Email softwares like Lotus Notes or MS Outlook and implementation of instant messaging service such as Lotus Notes Sametime Connect or MS Lync are examples of automation of communication. In addition, implementation of EPABX (Electronic Private Automated Branch Exchange) for a telephone system with voice mailbox helps in a higher productivity of the employees.

Office Resource Management

In large corporations, resources such as conference rooms, presentation tools such as projectors, etc. are always in demand. Automated tools help in managing these resources for effective use and for avoiding conflict in demand through advance booking. Through mail systems such as MS Outlook or Lotus Notes, the calendar and resource functionality can be leveraged to manage office resources.

Document Creation

In an office, several ad hoc documents have to be prepared based on needs. This could be a formal report or statements that have numerical calculations such as trend-analysis statement or a slide-presentation document. Microsoft Office suite and Lotus Symphony or Open office (open-source software) are useful for this type of automation.

Document Management

Documents created in a business enterprise need to be stored and maintained for various requirements. Certain documents have to be maintained based on regulatory requirements. For example, financial records of a company have to be maintained for about eight years under The Indian Companies Act, 2013. The important documents may also be retained for building knowledge within the company. Nowadays, maintaining the documents in digital form helps in effective maintenance of the document and it is permissible in certain laws also. There are automated systems that help in managing the document life cycle. These are called document management systems. IBM Filenet, Documentum are examples of document management system software.

Knowledge Management System

There are multiple environmental dynamics that demand a simultaneous continual learning for people in an enterprise. The advent of information and Internet technologies facilitated the making of the globe as a village; a concept of competitive advantages of nations emerged. Under this concept of competitive advantages of nations, the latter source that function from a country in which the country is having a natural expertise. As a result, the enterprises have started restructuring their organisations, moving different functions and operations to different nations. For example, IT-enabled services are the natural expertise of India and Philippines; these functions of most of the organisations have moved to these countries. Similarly, manufacturing has been moved to China, South Korea and Taiwan. When these kinds of changes are happening, there is a strong need for enterprises to relocate a functional operation from one geographical area to another; it calls for training new human resources freshly inducted in the new geography. Even though the human resources are moving out of the enterprise, the knowledge should remain. Thus, the Knowledge Management System was evolved.

Knowledge management is the process of capturing, developing, sharing and effectively using organisational knowledge. The concept of knowledge management was powered by information technologies for an enterprise to perform the activities relating to knowledge management.

Types of Knowledge

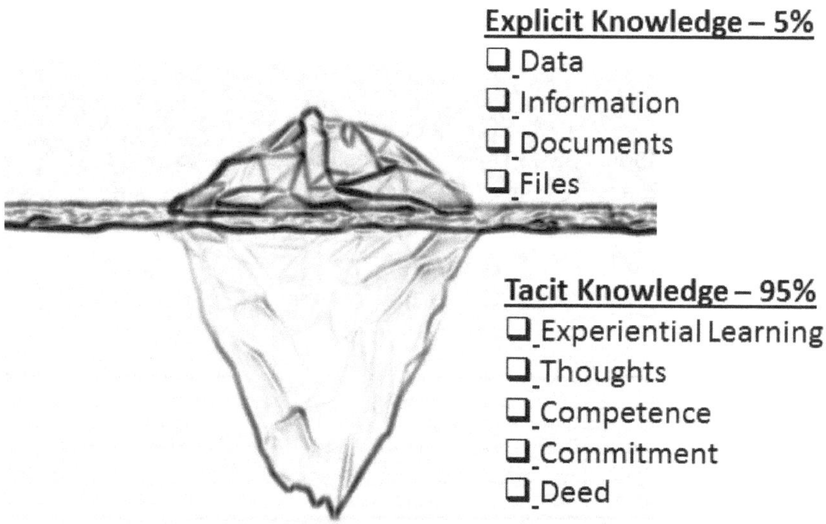

Diagram 6.8 Iceberg Metaphor: Explicit and Tacit Knowledge

There are two types of knowledge: ***Explicit*** and ***Tacit.***

Explicit Knowledge is the knowledge that can be formalised, codified, captured in documents, articulated and relatively easily imparted as a part of the knowledge transfer exercise. Examples of this knowledge are textbooks, user manuals, procedure manuals, documented standard operating procedures, etc.

Tacit Knowledge is in the mind of people—often in the mind of one individual—not captured by the enterprise knowledge management system. It is also difficult to capture tacit knowledge as part of a knowledge management system, as this knowledge is acquired over years of practice and experience by individuals and it is seldom possible to transfer tacit knowledge to another person who is inexperienced.

An enterprise, which is able to capture and retain tacit knowledge, gains a competitive advantage over the rest of the competitors.

From this perspective, an IT-based knowledge management system comes to help.

Enterprises create knowledge repository as part of intranet deployment. Also, enterprises encourage employees to participate in authoring and adding content to such knowledge repository and create a wealth of enterprise knowledge.

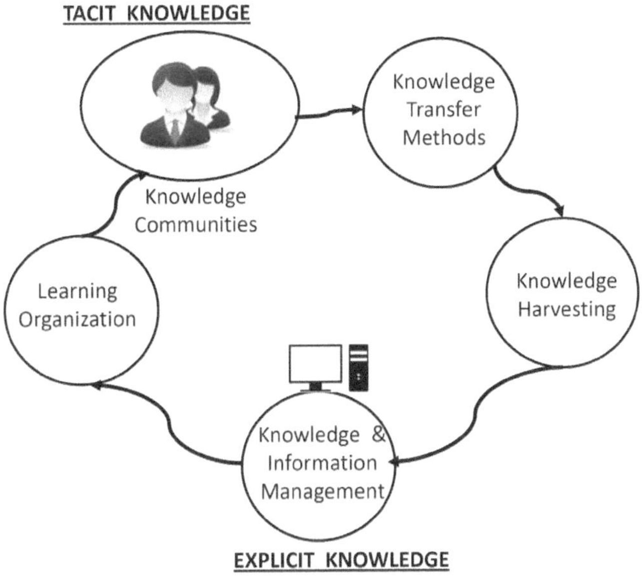

Diagram 6.9 Knowledge Transfer Process of Explicit and Tacit Knowledge

In order to ensure that tacit knowledge is also passed on to the next-generation employees, enterprises enable social networking and community building to enable the next-generation professionals to get mentorship from experts who possess the tacit knowledge.

Explicit knowledge is transferred, harvested, documented and stored in a knowledge repository. **Tacit** knowledge is provided through formal enabling such as mentorship, community interaction and socialisation.

1. Information and Knowledge

Information is said to be facts provided or learnt about someone or something. Information could also be false. Information is piecemeal and lacks connectedness.

Knowledge connects the fragments of information together. Knowledge forms the meaning from pieces of information. It brings information fragments together for gaining a wholesome idea and understanding.

Knowledge is more than loads of facts and information. Information is time-specific, whereas knowledge is generally not time-specific. Information flows, whereas knowledge is stock and static. Information flows and gets added to knowledge.

Information can exist without knowledge, but knowledge cannot exist without information.

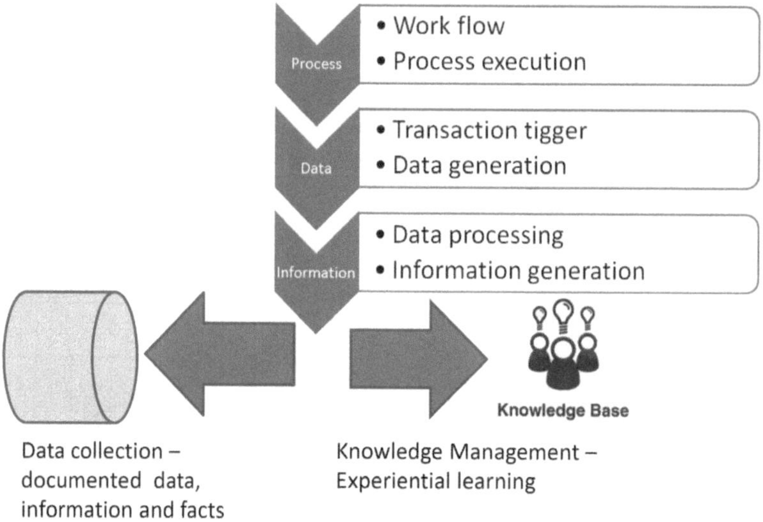

Diagram 6.10 Knowledge-Harvesting Process

Let us take an example to explain data, information and knowledge.

Assume we get a data that price is Rs1150—this is data. We get more information: the current price of a share of a company is Rs1150—this is information. We also have another piece of information that if a share price hits below a certain limit, it has to be immediately sold as per the company's stop-loss policy. We also have yet another piece of information: the price quoted for the share is held by our part of the company's investment portfolio. There are pieces of information. If an investment portfolio manager recognises that if the given share price hits Rs.1100 we have to immediately sell the share as per the company's stop-loss policy—this is knowledge.

2. Importance of Knowledge Management System

While information is power, applied information is more powerful. In order to use the information and to suitably apply it, one has to have knowledge. An enterprise's success depends upon its human resources that successfully apply information using their knowledge. As we have seen, tacit knowledge is gained from experience and experience comes from practice. Therefore, a knowledge management system is needed for the following reasons:

1. In a fast-paced changing business world, retaining the experiential learning of human resources is essential.
2. There is a need for replicating uniform knowledge among human resources across the globe because of rapid expansion and interconnectivity of enterprises in different parts of the world.

3. Globalisation has been adopted by many enterprises for economical sourcing and for expanding markets. This calls for a uniform knowledge about products and services.
4. An enterprise expansion calls for managing the enterprise expansion activities as projects. As strategic initiatives of expansion get over, people move from one project to another and there is a need for knowledge harvesting and archiving for future learning and reference.

Difference between Information and Knowledge

Information	Knowledge
Information is piecemeal	Knowledge is collection
Comes from data and fact	Comes from practice and experience
It is about 'What?'	It is about 'How?'
Information can exist without knowledge	Knowledge cannot exist without information
Information is time-specific	Knowledge is not time-specific
Information is added to knowledge	Knowledge receives information

Management Information System

A management information system provides information about business enterprise operations, provided to the management for an efficient and effective management of the business enterprise.

A management information system can be defined as the integrated elements of people, data, information and machine working together for providing information to the support functions and the management decision-making functions of a business enterprise.

The objective of a management information system is to provide all the necessary information to the managers of a business enterprise.

MIS: An Integrated System

In most of the enterprises, the MIS will be centralised as it provides services to all the business functions and often the management information requires a correlation of information between different functions, when producing the management information for serving the information service to the strategic management level.

The following are the phases of setting up a management information system:

1. "As is" study
2. Requirement gathering
3. Preparing blueprint
4. Implementation

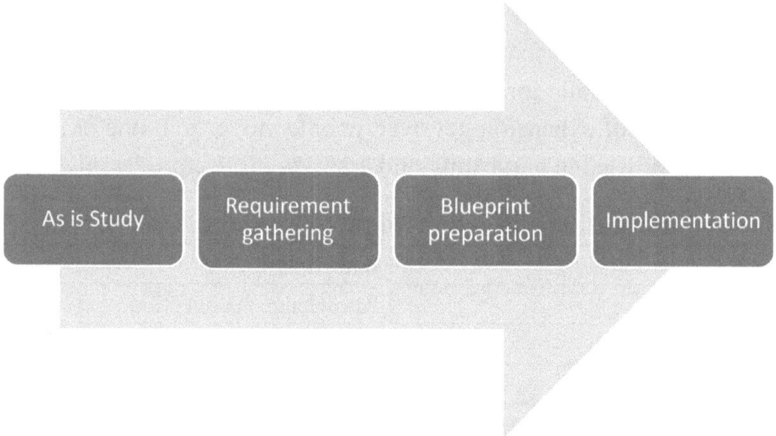

Diagram 6.11 Phases of Setting Up MIS

"As is" Study

This involves the following activities:

A. Studying the existing current system that provides information to the management
B. Limitations, drawback identification
C. Alternative approaches
D. Defining broad scope of a new MIS system
E. Cost–benefit analysis
F. Preparing a project report

Requirement Gathering

This step involves the following:

A. Identifying key users
B. Understanding the users' needs
C. Determining sources of data and information
D. Developing a high-level design of the MIS system

Preparing Blueprint:

This phase includes the following activities:

A. For each information need, there is the determination of the following:
   a. Inputs
   b. Processing method
   c. Outputs
B. Storage methods
C. Report design
D. Human resources and other resources' planning for implementation

Implementation

Implementation calls for the following activities:

A. Selection of hardware and software that are meeting the following criteria:
   a. Compatibility with existing systems
   b. Cost effectiveness
   c. Meeting the desired performance standards
   d. Support service
   e. Configurability
   f. Portability

B. Procurement of hardware and software
C. Installation of hardware and software
D. Coding, development and configuration
E. Training users
F. Preparing for go live (system launch)
G. Launch system (go live)
H. Maintain the system

Best Practices in Setting Up MIS

1. Keep the system as simple as possible so that it can be intuitive and can be easily understood by the layman. Avoid being ambitious.
2. Bring people together—users, management, IT organisation people—and come to a common understanding in terms of the needs and methods of development. Avoid an unrealistic development plan.
3. Involve developers and programmers from the beginning to enable them to understand the functional requirement and to decide the software limitation and explain in case of any unrealistic expectations from users.
4. Explore off-the-shelf—be ready to deploy software packages and customise to the enterprise needs, instead of developing from scratch.
5. Select simple software, instead of technically complicated ones.
6. Periodically review the progress with all to ensure that the solution developed is in line with the needs of the user and set right the expectations among all.
7. Implement in a phased manner module-by-module, instead of developing very large applications that become unmanageable.

Examples of MIS

- Passenger manifest produced by railway reservation system or an airline reservation system
- Account statements produced by core banking system for customers.
- Report generated out of ERP.

Decision Support Systems

A decision support system leverages data—historical and current—available in a transaction processing system and allows the decision makers to customise the information need for decision-making. The DSS is extremely beneficial to managers for the organisation's decision-making process. It supports the logical conclusions and basis for decision-making.

It enables the decision maker to view the information from multiple perspectives and different points of view.

In spite of having a decision support system, due to their incorrect interpretation and judgement, managers' decision could still be wrong.

DSS contains different tools and technologies. The new concept of 'Online Analytical Processing (OLAP)' systems are newer versions of the DSS system.

Components of DSS:

- Users/decision makers who use the decision support systems.
- One or more databases that contain data and information.
- Querying, programming language or customisable software interface.
- Model database/data mart/data store that stores the archived historical data.

Phases of Implementation of a DSS:

Unlike the implementation of other systems, DSS needs are felt by the enterprise as it grows, as the operations become more professional and as the senior management provides a complete autonomy to the junior managers for day-to-day operations following the strategic guidelines. The phases of DSS implementation are

Problem Discovery: As the enterprise grows, the need for DSS will be recognised and problems of not having a DSS will be discovered. As the senior management does not participate in day-to-day operations, it will not be in a position to take up a "judgement call" on the decision situation and will relay up on data, information and facts.

Solution Discovery: This is the next phase in which the business enterprise decides to set up a DSS. The enterprise explores options before it for setting up the DSS.

Solution Selection: After considering the various factors, finally, the enterprise decides to go with a particular solution selected from among the choices identified during the solution-discovery phase.

Solution Implementation: The enterprise sets up a project to implement the selected DSS solution.

Executive Information System

Executive Information System (EIS) is an infrastructure that supplies upto-minute information sifted from various sources, which helps the executives in their business decisions.

EIS is different from a decision support system. While DSS is for middle management, EIS is for senior management.

The senior executives need a summary of various critical business information and not detailed information.

Also, senior executives need information both from internal and external sources, as they need support to monitor the external environmental dynamics.

An EIS aggregates information from all other internal systems and presents a summary status of the business operation parameters.

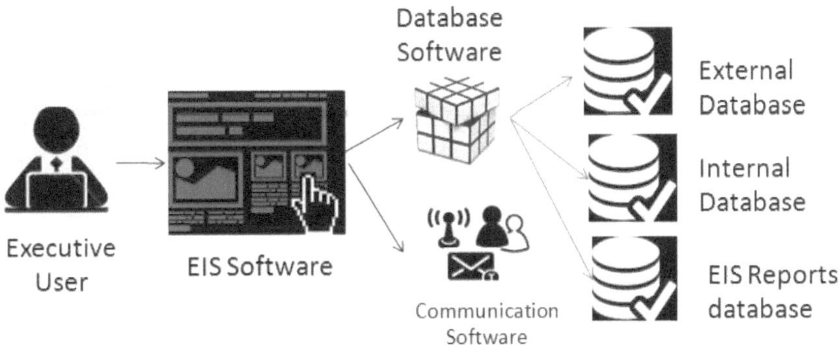

Diagram 6.12 Components of Executive Information System

Components of EIS

The components of EIS are hardware, software, user interface, communication software and databases.

Hardware: This includes all the standard high-power stand-alone personal computing laptop, docking station, tablets, mobile/smartphone device connections.

Software: All user-friendly softwares that aid the executive information system such as reporting software, graphic software, etc. that can present the various graphs.

User Interface: Interfaces for both standard computers and mobile devices with menu-driven commands with touch interface.

Communication Software: All communication softwares such as mail software, instant messaging, VoIP communication software.

Databases: Access to external databases (based on subscriptions), internal databases, business intelligence and analytical databases.

6.7 Specialised Systems

A specialised system provides comprehensive and end-to-end functionality features required for a specialised business operation. For example, petroleum pricing in the US for retail market is a specialised function and there are softwares that exist to assist business in the oil industry downstream. Similarly for the oil industry's midstream, operations like refineries, the laboratory information management system (also called as LIMS) system is a specialised system. Similar packaged software products are available for various industry specialisation requirements such as banking, insurance, pharmaceuticals, retail, telecommunication, media, entertainment, hi-tech, automobile, etc.

Enterprise Resource Planning

An Enterprise Resource Planning system is a packaged software. It is also called as ERP.

> An ERP is a software that integrates the internal and external management information of an enterprise taking on financial, supply chain, manufacturing, customer-relationship management and human resource management systems.

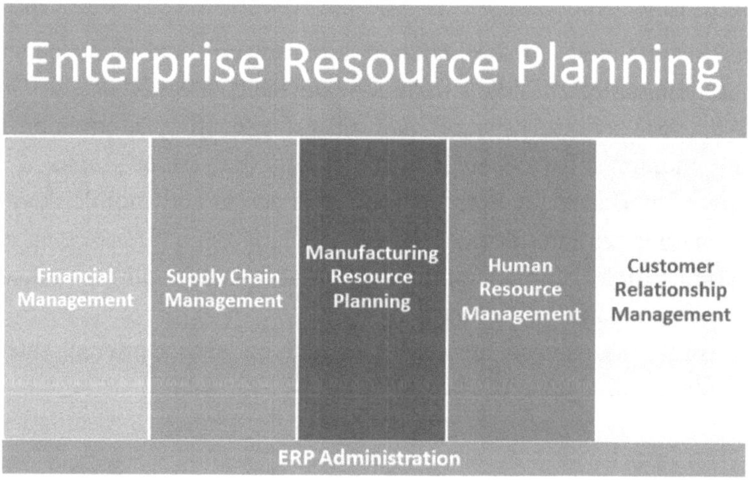

Diagram 6.13 Integrated ERP System

History of ERP Evolution

Specialised systems started evolving in the 1960s and evolved over the years. The pictorial representation of the evolution of ERP is depicted below:

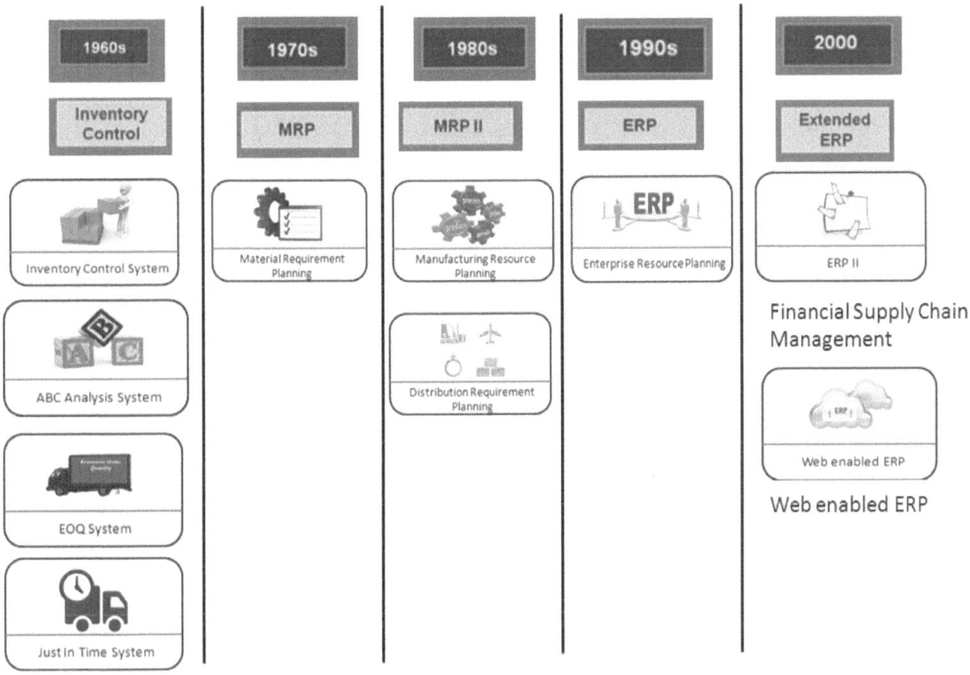

Diagram 6.14 History and Evolution of ERP System

1. 1960s was the decade when the industries witnessed an evolution of the inventory-management related software. It was realised that in any business about 40%–55% of the cost is towards the material ones and even if a fraction of a per cent is saved, that will save a huge amount of savings on the absolute basis. This era witnessed systems like inventory control, ABC analysis, Economic Ordering Quantity and Just-in-time applications.
2. 1970s was the decade when the PC revolution happened and every department in an enterprise had a PC. The material departments across industries attempted to further optimise the material cost through proper planning and by avoiding non-moving inventories. Thus, material requirement planning emerged. It is also called as MRP.
3. 1980s were the years in which manufacturing optimisation happened as the automobile industry in the US was losing to the Japanese car manufacturers in its domestic market. There was a strong need for optimising the cost of production, which demanded a meticulous planning of the manufacturing activities. Thus, the evolution of the Manufacturing Resource Planning—MRP II.
4. Around the same time, when MRP II was evolving, the distribution planning concept also emerged as a critical one and that resulted in an evolution of the Distribution Resource Planning or DRP.

5. In the1990s, information technology advanced with networking and all enterprises started adopting networking in their systems. Then the concept of system integration came up. In the same decade, the concept of Business Process Reengineering was propagated by Michael Hammer. As we know, BPR aims at the radicalisation of processes. BPR coupled with networking technologies in IT enabled the building of Enterprise Resource Planning software packages.
6. In the 2000s, ERP was further enhanced by the Financial Supply Chain Management functionality. Further enhancements such as variant configuration functionality (popularly known as CTQ or Configure to Quote) were set up. This was an era of ERP II. It was also called as Money Resource Planning (MRP III).
7. In the current decade (2011 onwards), all pervasive Internet technologies discovered new ways of computing over the Internet and new business models were evolved such as cloud computing, mobile computing etc. ERP progress witnessed web enabling. Now in an enterprise, ERPs can be accessed over the web through VPN and Internet. Besides, the ERP software is offered as a service over the Internet. For example, Net Suite is an ERP software offering. As of writing of this book SAP has announced its solution offering through cloud with their latest release called SAP S/4 HANA.

An ERP is founded on BPR with the power of networking technologies—it was thus able to automate the end-to-end enterprise business processes and to integrate them. Thus, an ERP has the feature of configuring all the required processes for their operations:

1. Order to cash
2. Procure to pay
3. Plan to produce
4. Record to report
5. Hire to retire

An ERP is basically a transaction processing system with a certain reporting capability.

The leading ERP packages are SAP R/3 for large businesses and SAP Business One for small business, Oracle E-Business Suite and Microsoft Dynamics.

Customer Relationship Management

Customer is the main reason for a business enterprise to exist and is the main source of revenue and profit.

Relationship is the process of continual communication and interaction with the customer to understand the customer needs and preferences better.

Management is the process of addressing the customer expectations, needs and preferences by the enterprise.

Thus, CRM is not just a function of marketing but of the entire business enterprise.

A CRM can be defined as a solution that helps the enterprises in establishing a long-term relationship with existing customers, manage customer loyalty, develop brand value of the products and enterprise, implement marketing strategies, generate information for evolving client specific strategies, etc.

History of CRM

The concept of CRM evolved in the mid-1990s, when the management gurus gave statements like the 'customer is a sovereign and decision maker in the modern business' and the 'customer is always right and let business be always right about customer'.

Enterprises have also realised that it is economical to retain a customer than creating/ developing a new customer.

There came also the CRM practice. As the information technologies advanced, CRM software solutions were developed and launched. For example, Siebel software was considered as one of the most popular CRM softwares. Siebel was subsequently acquired by Oracle.

Typical functionality of a CRM solution is

1. Long-term customer relationship with the established and new customers
2. Customer loyalty
3. Brand value
4. Marketing strategies
5. Client-specific strategies
6. Catalogue Management
7. Campaign Management

Analytical CRM

CRM software application is basically an analytical processing tool, which collects the sales and customer data from an ERP to analyse customer activities for understanding the customers.

Based on the customer buying pattern and behaviour, the business enterprise attempts to understand customer expectations so as establish relationship with the customers through constant communication.

A CRM software can offer more analytical information such as a territory-wise sales analysis, product-wise sales analysis, high-value customers, etc.

Greenberg Definition

According to the definition of Greenberg, "CRM must start with business strategy, which drives transformation of business and influences business processes enabled by IT".

The meaning is that a business strategy which addresses the need, expectations and preference of a customer is the starting point of CRM which aligns the entire organisational resources and enables the designing of business processes, which include the customer relationship processes.

Paul Greenberg further says, "The goal is to make the customer happy enough to want to continue the relationship with the company".

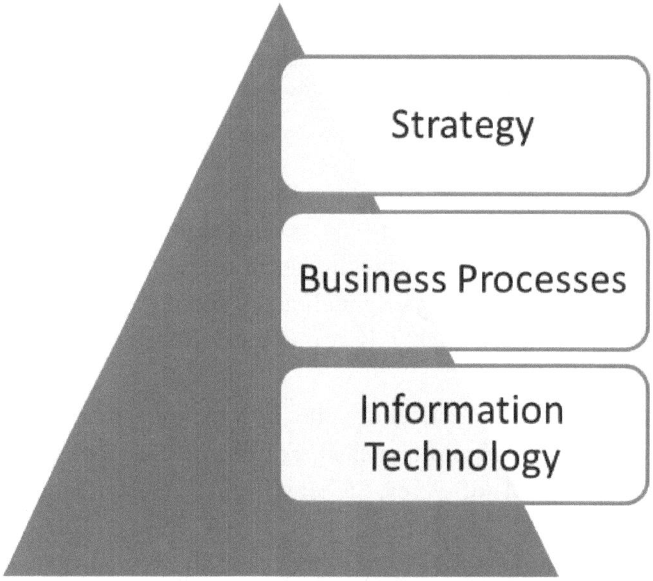

Diagram 6.15 Paul Greenberg's CRM Definition

Benefits of CRM

1. **Generates Customer Loyalty:** A continual relationship with the customers enables the enterprise to respond to the needs of the customers addressing their expectations and preferences. This helps the business enterprise to earn the trust of the customer and the trust lays the foundation for customer loyalty.
2. **Raising Market Intelligence:** As we have seen earlier, CRM is an analytical processing system and it helps the enterprise in gaining multiple perspectives of sales, market, product or service and customer hence raising the market intelligence through deeper insights.
3. **An Integrated Relationship with Customer:** CRM is an integrated approach to customer relationship, where each and every interaction with customers is captured for getting a better customer relationship.
4. **Preserving Existing Customer, Providing Enhanced Services:** CRM helps to retain the existing customers and through the careful understanding of customers, enables the provision of an enhanced customer service.
5. **Helps Entire Organisation:** CRM helps the entire enterprise to transform itself as customer-oriented and makes the process of capturing customer information and disseminating throughout the organisation smooth.
6. **Take All Action to Satisfy Customers:** CRM helps to initiate actions for satisfying the customer to ensure that the customer is happy and is retained by the enterprise.

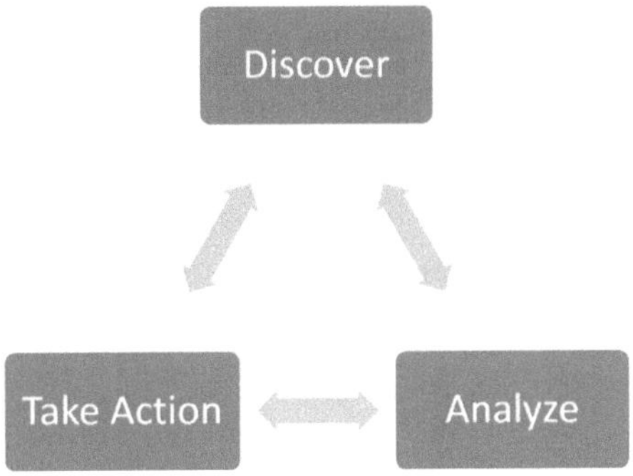

Diagram 6.16 CRM Process cycle

Why is CRM Needed in an Enterprise?

- It is the process for capturing customer information.
- Enables the enterprise to understand a customer.
- Enables a high-quality relationship as the enterprise is able to collect information on customer preferences,
- Data and information generated from a CRM system helps evolve business strategy.

Supply-Chain Management

Supply-chain management (SCM) is the management of the flow of goods. It includes the movement and storage of raw materials, work-in-process inventory and finished goods from the point of origin to the point of consumption.

Thus, SCM is an integrated approach to planning, implementing and controlling the flow of information, materials and services from raw materials and component suppliers through the manufacturing of the finished product for an ultimate distribution to the end customer.

Elements of Supply Chain Management

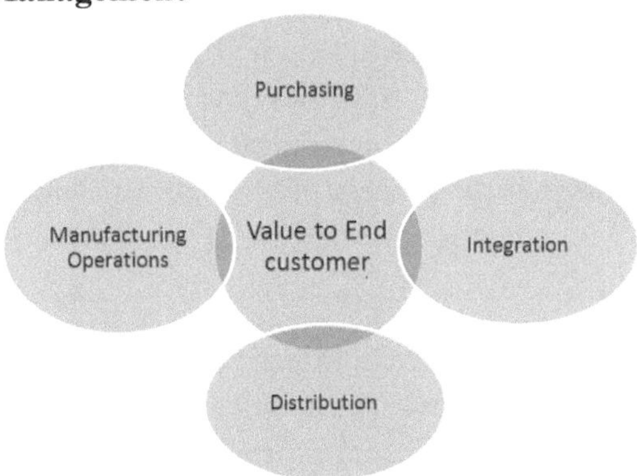

Diagram 6.17 Elements of Supply Chain Management

The four elements of supply chain management are 1. Purchasing 2. Operations 3. Distribution and 4. Integration

Purchasing

It is about supplier alliances, supplier management and strategic sourcing. Supply-chain management envisages a long-term relationship with suppliers.

The supplier management for an improved performance can be achieved through the following practices:

a. Supplier evaluation (determining supplier capabilities and performance)
b. Supplier certification (third-party or internal certification to assure product quality and service compliance to certain predefined standards)

Under SCM, the purchasing function aims at strategic partnerships for mutual success through the mutual trust achieved from long-term relationships with top-performing suppliers.

Operations

SCM operations handle demand management, MRP, ERP, JIT, TQM, etc. through the adoption of best practices.

SCM operations' 'demand management' function matches demand to the available capacity. SCM Operations link buyers and suppliers via MRP and ERP systems, use JIT to improve the "pull" of materials to reduce inventory levels and deploy TQM to improve quality compliance among the buyers and suppliers.

Distribution

The distribution element of the SCM handles transportation management, customer-relationship management, network design and service response logistics management aspects.

Transportation management calls for an optimisation of transportation operations by achieving well-defined trade-off decisions between the cost and timing of delivery (impacting customer service) via trucks, rail, water and air, without compromising on customer expectations. Customer-relationship management strategies ensure that right expectations are set about deliveries, complaints are resolved, communications are improved and service requirements determined. Network design is the process of creating distribution networks based on trade-off decisions between the cost and the sophistication of the distribution system.

Integration

SCM's 'integration' element handles coordination and integration activities, addresses global integration problems and performance measurement to achieve performance targets.

Supply-chain integration facilitates inter-firm functional integration when the supply-chain participants work for common goals. Supply-chain integration aims to transform attitudes and avoid adversarial relationships.

Supply-chain integration helps in setting up global supply chains. The advantages from larger global markets are lower costs and higher quality suppliers. It may involve certain operational exposures also.

The supply-chain integration helps in assessing the effectiveness of processes and procedures through performance measurement. This helps in deciding which process is working and which is not; when they fail to achieve the performance targets.

Relationship between ERP, SCM and CRM

Enterprise resources planning (ERP) enables the streamlining and management of end-to-end business processes.

Supply-chain management (SCM) helps in managing the chain of activities that involves the provision of a valuable product or service to end-customers, managing supplier relationship and logistics.

Customer Relationship Management (CRM) facilitates an enterprise to be in communication with the customer and to retain the customer with loyalty by meeting customer expectations and preferences.

All these are within the context of enterprise activities and help the enterprise in functioning efficiently and effectively to achieve the business goals.

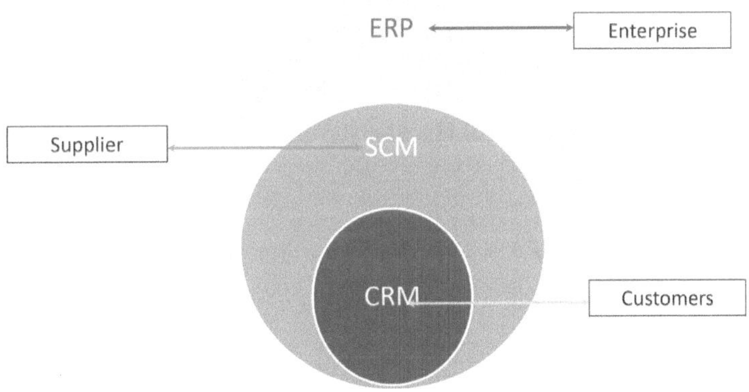

Diagram 6.18 Relationship among an ERP, SCM and CRM

Human Resources Management System

Human resources is considered to be the most valuable asset of any enterprise. It is human intellect through which business models, innovation, differentiation is conceived and makes the enterprise successful in the marketplace, in spite of competition.

Human resource management is also called as Human Capital Management. The functions have subfunctions of talent search, talent acquisition, recruitment, orientation, development, retention, progression and separation. Separation includes superannuation.

Human Resource Management System is a software (also called as HRMS software) that takes care of the processes relating to human capital management.

HRMS software will have several specialised modules. Refer to the diagram given below:

Diagram 6.19 Modules of Human Resources Management System

Workforce Management

Workforce management module of HRMS helps an enterprise to manage the human resources complying with the regulatory requirements and enterprise policy and procedure. It also helps in managing and optimising labour costs.

Time Management

Time management module helps all employees to log their time invested in various activities that helps the enterprise to serve the customers and achieve profitability.

Payroll

Payroll module helps the enterprise to process payroll and effect the payment of compensation to the employees as per the contract of employment.

Training

Training modules help to keep a track of the orientation and development activities planned for every employee. The programmes that have been undergone by each employee as as part of career development and in getting ready for the next higher role, are recorded and tracked in this module.

Compensation Management
Compensation management is one of the most critical aspects of human capital management. Evaluating the performance of each employee, deciding compensation enhancement that reflects the performance recognition and also fluctuation in the cost of living are the key attributes of compensation management. This module helps an enterprise in managing the compensation of its employees.

Recruitment
In certain industries like IT and IT-enabled services, recruitment is a process, whereas in other industries like manufacturing recruitment is an event. In either case, a recruitment module helps an enterprise to manage its recruitment-related activities.

Personnel Management
Personnel management involves the creation and maintenance of the personal records of employees. On most of the occasions, this is sensitive information. This module enables an enterprise to maintain all the information relating to every personnel.

Employee Self-Service
An enterprise continually communicates, transacts and interacts with employees. Employee self-service module enables an employee to help oneself with one's requirements according to the routine standard processes. For example, leave management is managed through the employee self-service. Similarly, submitting tax declaration and evidence for savings for income-tax exemptions, taking one's own salary statement from system, etc. are part of ESS.

Analytics
Analytics are a kind of business intelligence services offered by the HRMS system. For example, the employee turnover or attrition within the enterprise can be measured and can be compared with the industry benchmark. Similarly, the costs of labour can be compared and managed.

In addition to the above modules, certain HRMS systems provide additional modules based on the industry needs. For example, **Organisational development** is a module that enables the enterprise to define the roles and responsibilities of each category of employees. Banding has also enabled to group line-and-staff function employees to group based on the same years of experience.

CORE Banking System (CBS)
CORE is an acronym for **'Centralised Online Realtime Environment'** and as the name suggests, it is an online real-time transaction processing system for banks.

The CORE banking system is a specialised system for handling the basic processes in a bank such as retail banking, corporate banking, capital-market operations on behalf of a customer, wealth management and the cross-selling of insurance products.

Being an online and real-time system, it is available for customer operations 24/7. Therefore, CORE banking is integrated with the different channels of banking such as branch banking, Automatic Teller Machine (ATM), Internet banking, phone banking and mobile banking.

Reference functional architecture of a CORE banking system is provided in the diagram below:

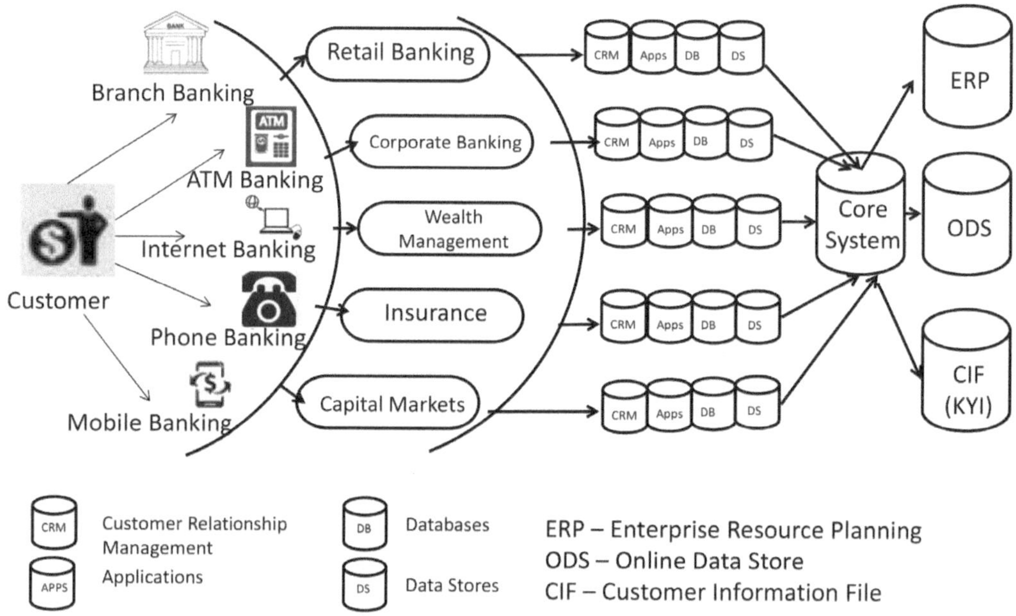

Diagram 6.20 Reference Functional Architecture of a CORE Banking System

Some of the CORE banking software packages that are offered by different software vendors are detailed below:

Software Vendor	CORE Banking Software Package
Infosys	Finacle
Nucleus	Finnone
Oracle	FLEXCUBE

Infosys Finacle

Finacle is one of the most popular CORE banking systems that is widely used by many banks across the globe. Functional architecture of the Finacle front-end is provided in the diagram below:

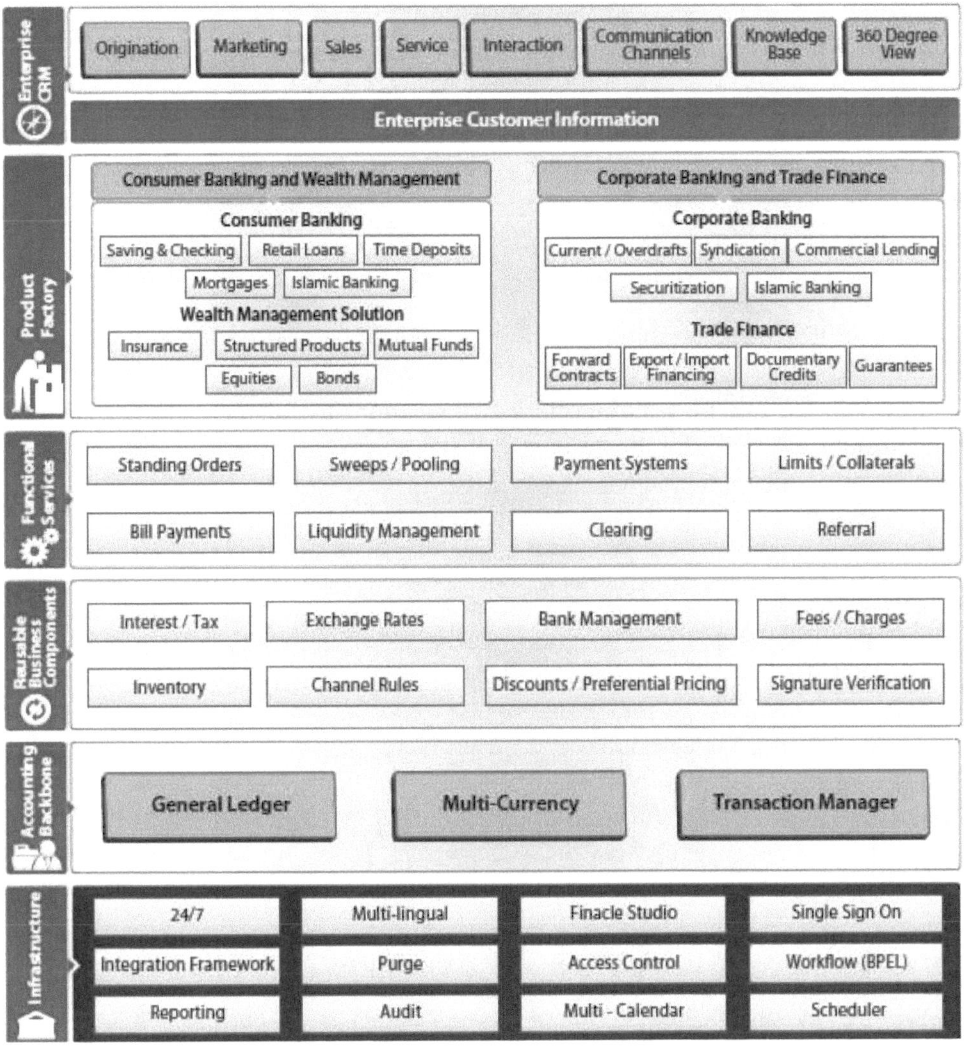

Diagram 6.21 Functional Architecture of Finacle

The key modules of Finacle are listed below:

1. Enterprise Customer Information System: All the key data relating to customers is saved in this module (In the Indian context this is known as 'Know Your Customer').
2. Consumer Finance: Retail finance such as home appliances, etc.
3. Consumer analytics: Analysis of consumers for decision-making.

4. Clari5: Enables the institutionalisation of customer relationship.
5. Wealth Management: Provides advice to customers and manages wealth management products.
6. Islamic Banking: Based on the needs and principles of the Islamic religious practices.
7. Corporate Banking: For corporate banking needs.
8. Trade Finance: Short-term working capital loan products for small business.

In addition, certain utilities offered along with Finacle are

- Payment lifecycle management across different types of banking
- Dashboards
- Organisation credit life cycles across different types of banking

Nucleus FinnOne

FinnOne is a CORE banking solution offered by the India-based software enterprise called Nucleus Software.

FinnOne solution is aiming at the offering of solutions to commercial banks and financial service companies.

Functionalities offered by FinnOne are

1. Retail banking
2. Corporate banking
3. Cash management
4. Relationship banking
5. Financial CRM
6. Credit Risk & appraisal
7. Enterprise application integration
8. Internet banking
9. Foreign exchange
10. Basel II
11. Data warehousing
12. Analytics

Oracle FLEXCUBE

FLEXCUBE was a CORE banking product developed by another Indian company called I-Flex. Subsequently I-Flex was acquired by Oracle and thus became Oracle FLEXCUBE.

FLEXCUBE houses almost all the standard functionalities and modules of a CORE banking software.

Besides, Oracle FLEXCUBE claims to offer universal banks with

1. Superior web experience with self-service and assisted channels
2. Improved bank staff productivity with intuitive role-based dashboards

3. Comprehensive transaction banking capabilities with the unique concept of Straight-Through-Processing
4. Improved risk management reporting

Accounting Information System

An accounting information system is a software that captures and records the accounting of data and processes the data for preparing financial statements. Through its different modules, AIS helps in providing more information for different types of management activities such as receivable management, cash-budget management, budgetary control etc.

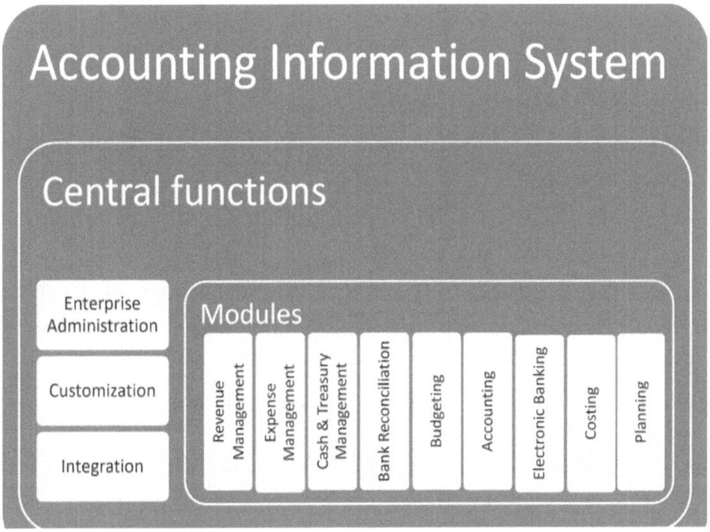

Diagram 6.22 Various modules of an Accounting Information System

Components of an AIS

The following are the elements of an AIS:

1. **People**: Various people such as users trigger business transactions and consequently accounting transactions get triggered. Besides the data generation that is captured by the system, the users include those who use the AIS for various purposes such as accountants, managers, auditors, consultants and regulatory agency representatives and they are covered under 'People'.
2. **Policies and procedures:** Policies and procedures of the enterprise are also a component. Based on the policies and procedures, the system is configured to ensure that accounting policies get implemented while automating the accounting transaction recording.
3. **Data:** Accounting and other data generated are obvious components.
4. **Software:** Software includes operating system, database and application software used for setting up the AIS.

5. **Hardware and IT Infrastructure:** Hardware and IT infrastructure such as servers, storage devices and network devices and software used for hosting and operating an AIS are the components of AIS.
6. **Internal Control:** Internal controls that are designed are a part of the AIS components. Unlike other functional softwares, in the AIS, the internal control plays a crucial role in view of its financial importance to the enterprise. Controls such as an expense claim submitted by a claimant reviewed by the claimant's manager or expense-payment approval at a higher level, needed for beyond a certain monetary limit, are some of the examples of internal control.

Functions of an AIS

AIS Generally Houses the Following Functionalities:

1. **Revenue Management**: Handles revenue recognition and recording. This may include receivable management also.
2. **Expense Management**: Handles expense accounting in conjunction with budgetary control. This may include payable management also.
3. **Cash and Treasury Management:** This helps an enterprise to plan for the sources of cash/ funds in the event of need and deploy surplus cash when the business has seasonal cash inflow beyond its needs.
4. **Bank Reconciliation**: Obtaining and recording bank account statements in the AIS, it automatically performs bank reconciliation.
5. **Budgeting:** Based on the business plan and the financial plan, the budgeting activity can be carried out with the help of the AIS.
6. **Accounting:** This means processing general ledger accounting through the AIS and in turn producing financial statements.
7. **Electronic Banking:** Through the banking module, an AIS system can affect an electronic funds transfer using certain predefined rules as needed by the financial system in the country.
8. **Costing:** In certain accounting systems this is called controlling; this module enables the enterprise to capture the conversion cost of manufacturing activities.
9. **Planning**: Based on the business planning and financial planning, a simulation is enabled by an AIS system.
10. The technical modules, are
    a. **System administration**: This is about administering the AIS system and setting up users, configuration etc.
    b. **Customisation**: This is about setting up the critical accounting processes that are required for an enterprise in the AIS.
    c. **Integration:** The AIS has to be integrated with the CRM, SCM and ERP for the benefit of the enterprise.

6.8 Expert System

An expert system is a computer system that emulates the decision-making ability of a human expert.

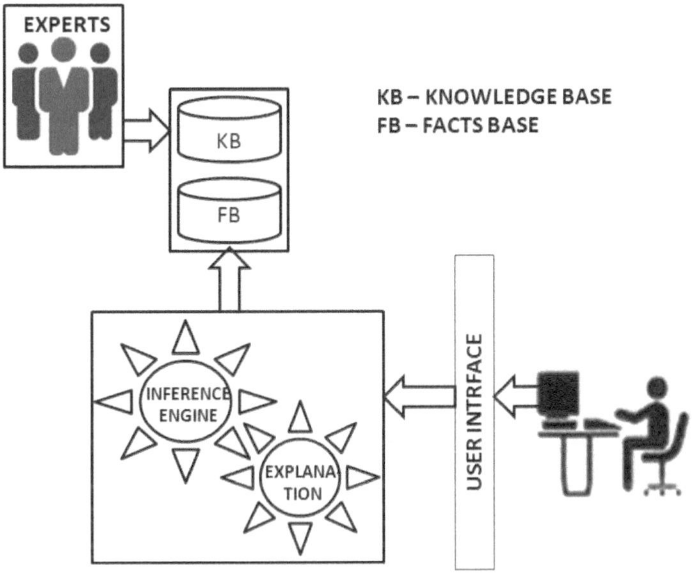

Diagram 6.27 Expert System

An expert system is part of Artificial Intelligence. By virtue of its capability, even if a non-expert uses the expert system, the system helps the user to take decision as an expert.

Components of the Expert System

An expert system will have the following components:

User: The user can be a layman or an expert; the expert system will respond to the queries of the user, irrespective of the expertise level.

User Interface: User interface typically provides the facility to design queries/questions and communicate with the expert system.

Interface Engine: Interface engine receives the question from the user and simulates expert logic processes and with the support of the knowledge base and facts database, responds with assistance to the user.

Explanation: Explanation is provided along with the output which explains why the expert system came up with the given recommendation. It explains the logical steps and the final conclusion arrived at.

Knowledge Database: The knowledge database is updated with the expert inputs with the knowledge on the given subject.

Facts Database: The facts database is filled with the facts based on the experience of the expert and the expertise on the subject.

Experts and Inputs: Experts are critical component of an expert system. Experts periodically provide inputs to the knowledge base and the database of facts. Their input is also a component of the expert system.

Types of Expert System

There are two types of expert systems based on the way of analysis and response to the user inputs.

Fuzzy Logic is one method, where the system handles imprecise data, by using level of confidence.

Neural Network is another type, where the system is designed to simulate the way in which the human brain processes the data/information.

Various Models of Expert System

Example-based expert system contains the cases, facts and results in the facts database and knowledge base in a 'T' form relationship. Facts of these cases are compared with the case on hand to provide the expert response.

Rule-based expert system is where the rules of the cases are stored in the form of 'If, then' formats and the system analyses the facts of the case on hand and applies these rules and provides the expert response.

In a **frame-based** expert system, the facts database and knowledge base contains frames. A frame can be defined as a data unit that contains data, description, rules, etc. Then, rules are defined on how to correlate these frames and assemble these frames to analyse a given case on hand.

6.9 Business Intelligence

Business Intelligence (BI) is a set of techniques and tools for the transformation of raw data into meaningful and useful information for business analysis purposes. Business Intelligence is much beyond the standard reports generated from the business application systems.

From a technological perspective, BI is an "online analytical processing system" (also known as OLAP). The standard systems such as ERP, SCM are to handle the transactions in a business. Therefore, these are called "online transaction processing systems".

Fundamentally, the data content of these two systems is different. In a transaction processing system, the data ia stored in a granular level for each transaction. This is valuable for a given period of time. Over the years, this data becomes very valuable for mining and taking out intelligent information out of it such as pattern analysis, forecasting future trends using statistical techniques, etc. An OLTP system cannot handle such complex analytical reports and one needs an OLAP system.

In an organisation, the transactions from OLTP are ***extracted*** and ***transformed*** in a particular fashion based on rules and ***loaded*** into the OLAP databases. This is popularly known as ETL process. The data stored in OLAP is called ***data stores, data mart or data warehouse.*** This transformed data is useful for producing analytical reports.

The diagrammatic view of ETL process is given below.

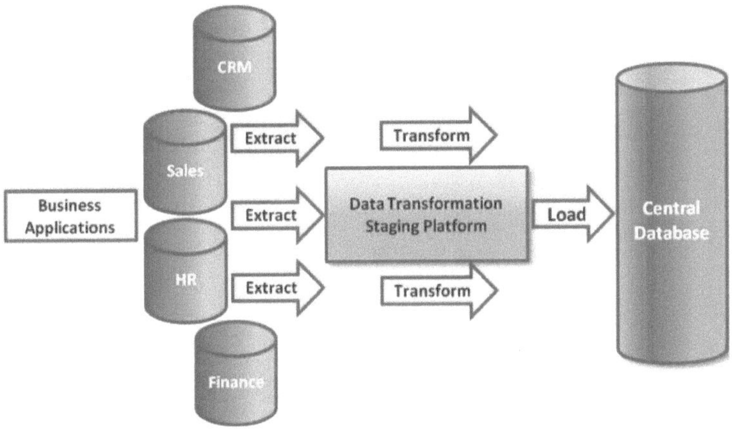

Diagram 6.28 ETL Process of a BI system

Information Reporting in Business Intelligence

There are multiple ways in which information is reported by the BI systems. These are

Simple Reporting by Querying

The user can have standard queries that can be executed from the user system; the BI system is invoked and response to the queries gets displayed on the user's terminal and if needed a printout can also be generated.

Dashboards

Dashboards are the reporting of the status of various business parameters in a single screen. For example, orders booked, inventory status, revenue, expenditure, receivable, payable, cash position—all this information can be displayed in a single screen through a customised program, which will connect to a data mart and fetch information and display it on the single screen. This data can be for a given period of time in comparison with the past period and also provides trending.

Various business decision-makers look at various operational parameters of their interest and use the dashboard for monitoring. For example, if the example given above is for a CFO, a Vice President in-charge of manufacturing may be interested in the data relating to manufacturing such as units produced, shipped, capacity utilisation of various machines in the plant, productivity of labour, manufacturing plans for the upcoming period, maintenance plans, etc. So as for the leaders of marketing, procurement, etc. Sophisticated dashboards are possible only with the help of a BI system and it is very complex to construct a dashboard. An example of a dashboard is given in the diagram below:

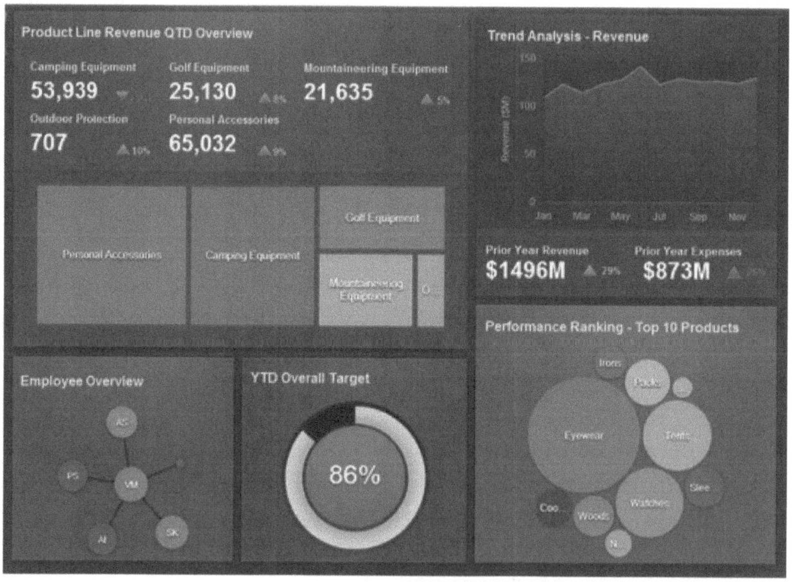

Diagram 6.29 Dashboard

Scorecards

In a scorecard, generally, the information on target to be achieved with respect to a parameter of a business is compared with the trending number against that parameter is provided. Scorecards are something similar to dashboards, from a technical perspective, where the data is extracted from data marts and provided to the user along with the target score for a user to compare and evaluate a parameter.

The best example of a scorecard is a 'Balanced Scorecard'. An example of a balanced scorecard is given below:

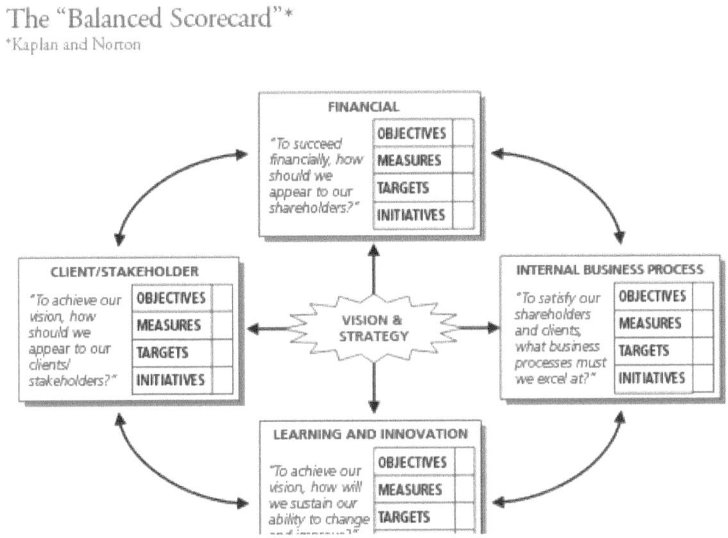

Diagram 6.30 Balanced Scorecard

Data Mining, Statistical Analysis, Business Analytical Reports

The above are generally interchangeably used. Business executives always tend to take a look at the data and facts before taking any decisions. From that perspective, often there will be a need for projections/predictions for the future from the past events and associated data. For this purpose, statistical methods such as correlation analysis, regression analysis, etc. are used.

Certain BI systems also enable operation research algorithm in programming to present reports for predictability management. All of the above are possible only through a BI system. BI has replaced the previous generation decision support system (DSS).

Section 7
Deploying Business Applications Software

7.1 Background

When a computer is applied for automating the business processes, it is done through a process called application development. These applications are generally called as application software. Business applications are a set of softwares used in business by people for conducting business.

Examples of business applications are accounting information system, inventory management system, cash management system, etc.

Types of Business Applications

There are several ways in which business applications can be classified. Here are some of the bases on which we can classify business applications:

A. Types Based on Nature of Processing

There are two ways in which we can classify business applications based on nature of processing as below:

Online Real-Time Processing

These are applications, where the business data are processed online in real-time.

Batch Processing

In this type, the data is collected and stored in the database by the system, but the processing will not take place in real time. The data will be processed in batches but once in a predefined periodicity—say, daily, weekly or monthly.

B. Types Based on Source of Applications

Applications in a business can be set up through the different sources as listed below:

In-house Development

In this case, the team from the IT organisation would have built the application for the exclusive use of the enterprise. In the 1950s and 1960s, large corporations using mainframe applications used to build their applications with their IT organisation personnel.

Custom Development

This type is something similar to that of the 'in-house development' software type but with a difference that instead of the software being developed by an in-house IT organisation, a third party is hired to develop these applications for the exclusive use of the enterprise.

Packaged Software

These are off-the-shelf products developed by the software product enterprises for a given industry. For example, Enterprise Resource Planning (ERP) software is a packaged software and there are different variants of ERP for each industry. For example, SAP IS Oil is exclusively for the petroleum industry and SAP IS Retail is for the retail industry.

Packaged software can be further classified into two more categories:

Licensed Software

In this case, the enterprise using packaged software has procured the license to use the software by its business users. The enterprise pays the license fee periodically. The licensed software will be deployed within the IT infrastructure of the enterprise that has procured the license.

Software-Usage Subscription

This is a new model, where the software is sold as a service. In this type, the provider has set up one's own infrastructure and deployed the software package and allows the customer to use the software as needed on a 'pay-per-use' basis. This type is also called cloud services or cloud computing as the customer accesses the packaged software in the provider's environment through an Internet cloud.

C. Types Based on Target Business Segment

Different types of software products are developed and sold to different business segments as below:

Home Users

Product developed for home users are simple and may not have all the features that the businesses need. For example, Microsoft Office Home premium falls into this category.

Micro-Business Applications

Micro businesses are different from small businesses. Micro business is a very small business that is run by a self-employed person from one's home. Mostly, these types of professionals use home-user applications for the business needs.

Another school of thought of micro business addresses or serves the micro business. For example, micro finance is a process of mobilising small savings from the remote parts of the country where the banking system does not exist. The process of collecting savings from the agricultural daily-wage labourers and providing access to banking is micro finance. Software application needs of a micro business are totally different from the needs of standard commercial banks.

Small and Medium Enterprise Business Applications

There are software packages that have been exclusively developed for the use of small and medium businesses. For example, SAP Business One is an ERP application that was developed and sold by SAP specifically for small and medium businesses.

Large Enterprise Applications

Application software developed and sold to large-scale enterprises fall under this category. An example is the SAP R/3 ECC6 software sold by SAP for large businesses.

D. Types Based on Functionality

There are different types of applications based on functionality and some of them are

1. Enterprise Applications: ERP, CRM, SCM, HRMS, etc.
2. Manufacturing Execution System: for managing manufacturing activities.
3. Laboratory Information Management Systems: for use in laboratories.
4. Retail software: for managing retail business operations.
5. Security software: for protecting an enterprise's IT security.
6. COREBanking System
7. Product Lifecycle Management System (PLM)
8. Legal Management software
9. Logistics management software
10. Industry-specific software

7.2 Business Process Automation

A business process is a series of activities through which inputs are processed and converted into output. Business processes are series of discrete repeatable tasks; tasks sequence results in work-flow and is performed by human beings.

Business process automation is automating the business processes, generally applying IT, to achieve the saving of cost, time and effort.

Automation of all activities in business process and work-flow is called Business Process Automation (BPA). An automated process can be simple or complex; irrespective of the complexity level, the following advantages can be derived from a business process automation:

a. Activities are carried out more speedily than in the manual method
b. Errors are minimised
c. Human efforts are engaged in a more valuable work

The following steps are involved in business process automation:

1. Defining the purpose of business process automation
2. Developing an understanding of the rules and the procedures to be complied with while automating the business process
3. Documentation of the process details to be automated
4. Detailing what are the goals proposed to be accomplished by automating the business process
5. Engaging a business process expert for review guidance
6. Preparing a project proposal for the BPA, including ROI details
7. Implementing the BPA project
8. Testing the automated business process
9. Going live with the automated business process environment
10. Carrying on with the ongoing continual improvements

Business Process Management

Business Process Management, often called BPM, is the methodology through which the enterprise processes are monitored and continually improved. BPA automates the processes of the business.

An Enterprise Resource Planning (ERP) often has BPA included in it and implementation of an ERP system and BPA involves the following phases, as covered in the subsequent sections.

7.3 System Development Life Cycle (SDLC)

System Development Life Cycle is also called as Software Development Life Cycle; in short, SDLC is a methodology used for the development of a software system.

SDLC involves five steps in developing a system or software:

1. Investigate
2. Analyse
3. Design
4. Implement
5. Maintain and review

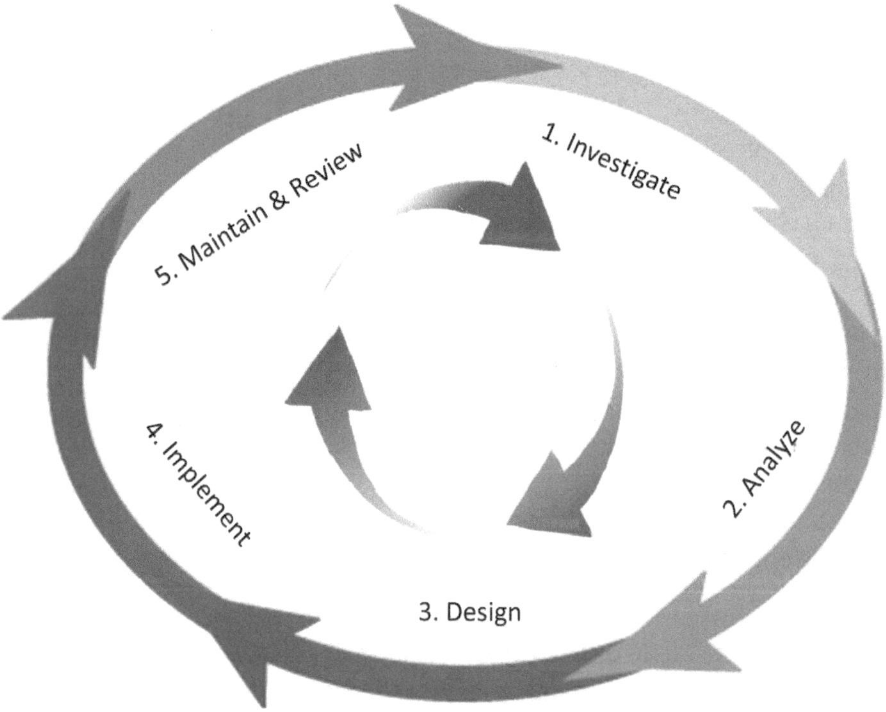

Diagram 2.14 System Development Life Cycle

"Investigate" Phase

During this phase, the problem or the challenge confronted by the business/enterprise is analysed. The needs of the users are gathered to understand the challenge and the problem. After defining the problem, the problem is evaluated in the light of the worthiness of resolution. Also, under this phase, various feasibilities are evaluated as below:

Technical Feasibility

The answer to the question that 'is there any technology available to resolve the defined problem or challenge?' is explored.

Economic Feasibility

How much would it cost to implement a solution to address the problem or challenge is evaluated and whether it is feasible to invest is also questioned in this phase.

Legal Feasibility

The issues of 'is there any regulatory requirement or legal aspect to be taken care of while addressing the issue', 'is it feasible under the current law to implement the solution to address the challenge/problem' are also evaluated in this phase.

Operational Feasibility

If a solution is implemented to resolve the challenge, is it practically possible to operate smoothly? This aspect is also evaluated.

Schedule Feasibility

Do we have sufficient time to address the issue, evolve a solution and implement? The answer to this is also considered at this phase.

Analyse Phase

System analysts typically perform the activity of analysis. System analysts interview people, study current processes, review standard operating procedures and gather additional information circulating questionnaires.

After gathering all the information, system analysts examine it and develop proposals for the new system features, perform a cost–benefit analysis, develop a business case, outline the system requirements and the hardware requirement during this phase.

Design Phase

In this phase, the actual solution is developed. The activities included in this phase are

- Determining the required IT environment with hardware, server, storage and software
- Developing designs and architecture of the solution
- Preparing detailed design documents incorporating input, output, user experience requirement, performance parameters, etc.
- Preparing a detailed test plan

"Implement" Phase

The newly designed solution is implemented in this phase and the new system goes live. The below-mentioned activities are also performed during this phase:

- Acquiring an IT environment with hardware, server, storage and software
- Code development
- Testing
- Go-live preparation

There are different methods of going live:

- Direct change over: This is also known as the big bang approach, where the old system is retired and all features of the new system become operational for the users.
- Parallel conversion: In this case, the new system and the old system will coexist for a period of time, till such time that the new system stabilises.
- Phased conversion: Under this method, the new system will go live in a phased manner and the old system's retirement also will be in a phased manner.
- Pilot conversion: In this approach, one part of the new system will be going live to evaluate whether the solution and technical aspects are in line with the expected solution. Once the pilot run is successful, the rest of the solutions will go live.

"Maintain and Review" Phase

During this phase, the newly gone-live system will be maintained and reviewed for further enhancements. Maintenance activities include the following:

✓ Resolving minor program issues
✓ Adjusting the operating procedure
✓ Making amendments to the program for better performance
✓ Enhancing the program with minor but improved functions

System maintenance could be:

Perfective maintenance: looking for an opportunity to improve the system

Adaptive maintenance: making the system adapt to the enterprise environment and the periodic maintenance cycle—for example, the system has to be upgraded with the latest version of the software to adapt to the current version of the software.

Corrective maintenance: Fixing issues as and when they arise is corrective maintenance.

Advantages of BPA

Automating business processes offers a variety of advantages as detailed below:

Reduction in Error: Automated business process execution helps in a significant reduction in human errors, as human efforts are involved only at the time of setting up the automated business process and wherever human response is needed when the process is under execution.

In other words, there is a significant reduction in human involvement, which in turn reduces the errors.

Transforming Data into Information in Real Time: As the business process automation handles business processes in real time, the business data gets converted into information in real time.

Speed in Process Execution: Automation helps in accelerating speed without errors; hence, the process becomes more efficient and more effective.

Making Human Resource of the Business More Efficient: As a result of the automation of business processes, the human resources released from the manual process execution are now available for more intellectual work than the mechanical repetitive work, thus making the human resource in business more efficient.

Making the Business More Responsive: As the business process automation helps to update the status of the process execution progress, people can determine the status in real time.

Enabling Collaboration and Information-Sharing among People: As business processes eventually cut across different functions, business process automation enables a collaboration among the people and helps them in sharing information.

Need for IT Infrastructure

In order to achieve the automated business process in a business, the following IT infrastructure is needed: hardware, network infrastructure, BPA software, databases and storage. The below-mentioned activities are carried out by the BPA, from an IT perspective:

- Database access, changes and updates
- Administration
    - Operating system administration
    - Database Administration
- File replication and back up
- System and event-log monitoring
- Job scheduling
- Application integration
- File transfers
- Printing

Automated Business Process Management

We have seen in the previous section as to how IT accelerated and strengthened the value of Business Process Management.

ERP packages coupled with BPM software have armed enterprises with a powerful way to manage business processes with more efficiency, effectiveness, conformity and compliance.

By applying IT through the usage of BPM software, enterprises have derived immense business benefits as below:

1. Better control over processes with robust authentication, authorisation and approval processes.
2. Overall cost reduction with higher efficiency.
3. Speedier processing of transactions.
4. Transaction processing and approval process are beyond time and geographical boundaries when applications are accessed by employees of the enterprise over web-based access through the Internet.
5. Enabled enterprises to quickly respond to changes in the business environment with a renewed business process implementation. For example, centralised back office.

Why BPA?

Cost Saving

Business process automation helps in setting up efficient and effective processes while completely avoiding human errors. Hence, it helps the enterprise in overall cost saving.

To be Competitive

By achieving cost saving, and by passing on that saving to the customers in the form of price reduction, an enterprise becomes more competitive in the market.

To Serve the Customer Faster and Better

Automated business processes help to speed up the process performance; thus, this enables the enterprise to serve the customers better. Besides, through business process automation, by setting up a self-service portal for customer service, customer service gets enhanced.

Objectives of BPA

Confidentiality

BPA—by virtue of its design—will allow only authorised persons to access the enterprise data and information; hence, confidentiality is achieved.

Integrity

As BPA achieves automation with a preset reference data and avoidance of human interference, it ensures that the data in the system is reliable with a high order of integrity.

Availability

In a BPA environment, the information availability is ensured at any time and almost instantaneously.

Timeliness

BPA ensures timely availability of data and information and often the information is available all the time (24/7).

Power and Value of BPA

To understand the power of automation, one can consider two examples:

Amazon.com

Amazon.com was founded in the year 1994; it achieved the quarterly revenue of US$12 billion in Quarter 4 of year 2012. As compared to Amazon, Barnes & Noble founded in 1886 (with the name Arthur Hinds & Company), a conventional bookstore, posted a revenue of US$1.1 billion revenue for the same period.

Flipkart.com

In India Flipkart.com, which was set up in 2007, has the higher business valuation than a leading retail chain—Futuregroup. The automated, online, e-Commerce retailer Flipkart is valued at Rs30,000 crores by Venture Capital as against the Futuregroup which is valued at less than Rs3000 crores. As most of us are aware, Flipkart has just 6 large warehouses across the country whereas the Futuregroup has about 319 retail units in different formats across the country.

Implementing BPA

Implementing a BPA is a project. Generally, these projects are taken as a part of strategic initiatives. Broadly, there are about 8 steps involved in a BPA project as listed below:

The above steps are explained in detail along with a case study:

Diagram 5.1 Steps in Implementing BPA

BPA Implementation Step Details	Case Study Events
Step 1: Investigating the Need for BPA The primary purpose and need for business process automation may vary from enterprise to enterprise. For certain enterprises, the need could be just cost saving; for another, it could be speed/high volume handling/error reduction/better customer relation, etc. Basically, the enterprise's current process needs to be reviewed and what is the challenge that is faced by the enterprise has to be identified to have a business case for recommending a BPA implementation	**A State-Owned Insurance Regulator (SOIR) in the USA** is a government department that was created in the last century as a part of a national system of state-based insurance regulation. The insurance marketplace has changed dramatically over time, but consumer protection continues to be the core of SOIR's mission. As of 2010, SOIR was responsible for regulating 1500 insurance companies and 300,000 insurance agents and brokers, with the budget allocation by the government at the same level as of 2005. With budget cuts, cost cut was compulsive. Besides, the agency mandate was to meet the needs of the stakeholders and the public better by establishing a foundation for a continuous process improvement. Need was (1) Cost reduction (2) Laying foundation for a continuous process improvement.
Step 2: Understand the Regulatory Requirements This step involves understanding the rules and regulations that the enterprise need to comply with such as • Legal requirements • Document retention requirement under law. • Policies and procedures of the enterprise, etc.	**Rules and Regulatory Requirements** SOIR issued approximately 1200 licenses annually to insurance companies and each licensing application had to be reviewed by SOIR attorneys, financial analysts and investigative staff residing in different locations across the West Coast. License approvals required documents to physically move between cities, forcing a serial review process as each department's staff handled tens of thousands of pages of paperwork. In addition, all insurers were required to submit, in paper, financial documents to SOIR, such as Annual and Quarterly financial statements, State supplemental filings and annual renewals. Payment fees were submitted through cheques, resulting in more manual processing.

Step 3: Documentation of the Process to be Automated Each and every process step details have to be reviewed, understood and documented, such as: What data and documents are to be captured? What are the sources of these documents and data? Who all are involved? What regulatory requirements are to be taken care of, etc.? The document should be clear so as it benefit every entity and individual participating in the BPA implementation programme.	**Software Services (SS) and BPM Experts: Partners of SOIR** SOIR collaborated with Software Services (SS) and BPM Experts to develop and deploy SANCTUARY, the Online Assistance System for Insurer Submittals, which facilitated the electronic processing of all corporate transactions. SS and BPM Experts completely documented the business process and regulatory needs and offered to design a configurable data model that allows for re-useable components supporting future services and additional applications.
Step 4: Define the Goals to be Achieved by Implementing BPA This step involved the definition of the key objectives of process improvement activities that are specific, measurable, achievable, realistic and time-bound.	**SOIR Goals for BPA** (1) Cost management (2) Lay foundation for a continuous improvement (3) Paperless office (4) Scalable BPA solution (5) Save Papers
Step 5: Engage a Business Process Consultant The main objectives of engaging a business process consultant is to ensure that (1) a subject matter expert is involved (2) an outsider view is gained by the enterprise (3) bring the best practices available in implementing the project are brought in.	**BPM Experts was the consulting partner for SOIR** which helped the SOIR in extending the necessary consulting, bringing subject matter expertise and project management expertise along with bringing in the best practices on BPA.
Step 6: Calculate ROI for the Project: In this step, typically, a cost case is developed and the investment is justified by the benefits. Every benefit—tangible and non-monetary—can be considered and assigned a monetary value to justify. For example, space saved by automation, by avoiding a manual record keeping, can be one of the benefits and a monetary value can be assigned to it. All the benefits with corresponding monetary values are considered to evaluate the investment worth in terms of how the benefits outgrow the investments.	**SOIR Anticipated the Following Benefits, though the Complete ROI Details are not Available:** 1. More than 1000 application filings annually 2. Manage about 25000 documents electronically every year 3. Save at least 1 million sheets of paper every year 4. Efficiency gain in terms of at least 2 personnel a year in the legal department alone

Step 7: Develop the BPA to be Implemented In this step the solution to be implemented is developed and deployed.	**SS and BPM Experts Jointly Developed and Implemented the Solution for SOIR** SS and BPM Experts jointly implemented SANCTUARY, the Online Assistance System for Insurer Submittals, which facilitated the electronic processing of all corporate transactions. SANCTUARY is powered by the Software Services (SS)'s Business Process Management (BPM) solution—SS Business Manager (SSBM)—which offered a configurable data model that allows for re-useable components supporting future services and additional applications.
Step 8: Testing the BPA In this step, the developed new system will be tested before launch to ensure that all requirements have been met and the new system is meeting the targeted solution.	**SOIR Users Tested the System before Going Live** SS and BPM Experts worked with SOIR to train internal users. SSBM solution converted paper-based application processes to an online system, allowing the insurers to upload filings electronically in a fraction of the time. SANCTUARY now acts as a unified portal that allows insurers to electronically submit applications, fees and other transactions on the SOIR website, by simply using the portal, instead of manually filling out an application in triplicate and mailing it in.

Applications that Help the Enterprises to Achieve BPA

Accounting Applications such as *Tally*

An accounting package helps an enterprise to automate the maintenance of books of accounts and finalisation of the statement of accounts. Besides, an accounting package helps an enterprise with various information reports that are very useful for management decision-making such as receivable reports. Most of the accounting software packages developed locally help the enterprises to maintain their records for the needs of statutory requirements such as tax compliance, income-tax law compliance, etc. Generally, an accounting application provides information from a summary level to the granular level, with a feature called 'Drill down', at various stages.

SAP R/3

SAP R/3 is one of the world-renowned Enterprise Resource Planning (ERP) softwares. SAP is the market leader in the ERP segment and has further expanded its software solutions to various industry needs through industry-specific solutions, functional analytic needs such as

SCM, CRM and technologically advanced solutions such as Business Intelligence, Advanced Planning Optimization (APO), etc. SAP R/3 has several functional modules, such as: financial and controlling—take care of accounting and costing, bank accounting, cash-budget management, funds management etc; Logistics management— takes care of procurement, material management, production planning, quality management, warehouse management, sales and distribution and HRMS—takes care of personnel management, time management, payroll, compensation management, travel management, etc.

MS Office Suite of Applications

Most of us are familiar with the MS Office Suite of applications that contains MS Word—a word processor for composing documents, MS Excel—a spreadsheet for mathematical and statistical calculations, MS Power Point—for preparing presentation documents. Certain versions of MS Office also include MS Outlook, an email communication software (client-end components).

MS Office is one of the most sought-after softwares for office automation, globally and widely popular. Not only enterprises but also individuals such as independent professionals, students and families are using MS Office. MS Office has been developed with a user-friendly intuitive interface that has made it possible for any individual to easily adapt to it.

Automated Access Control System

Nowadays, most of the entities such as business enterprises have installed an automated access control. There are two types of access control systems: (1) identity-card based (2) biometric based

Identity-Card Based Access Control System

In this case, all the authorised people, who have access to the facilities of the business enterprise, are provided with an individual identity card with a magnetic strip behind the card. There will be an access control device at the entrance of each facility. The individual who wants to gain access to the facility needs to either swipe or wave the card (in case of NFC technology) and the device captures all the information about the individual and verifies with a centralised server-based software to check whether the individual is authorised to access the facility or not. If authorised, the individual is allowed; otherwise not. Some institutions use this system for monitoring the attendance of the employees also.

Biometric Based Access System

The functionality of this type of system is identical to that of the identity-card based system, except that instead of using the identity card, the individual's fingerprints are stored in the system to identify the authorised individual. To gain access, individuals have to place the thumb and fingers on the sensor of the access control device for scanning fingerprints.

Vehicle Tracking System

Vehicle tracking systems are very popular systems and widely used for a variety of purposes. A transportation department of a private business enterprise can use this for tracking their trucks used for logistics purposes. A general insurance company uses this for capturing the information about the drivers' driving behaviour, based on the driving pattern and to predict risk and manage the insurance cover/premium accordingly. Safe drivers are offered higher discounts in premium and vice versa. Cab operators (enterprise) to track the movement of the cab can use a Geographical Positioning System (GPS) system. Nowadays, mobile phones with General Packet Radio Service (GPRS) along with map software meet the need of a GPS system. Since the GPRS is a connection based on the Internet facility available in a location provided by the telecom service provider, it may not be available, if there is no Internet service or a telecommunication service in a particular location. But, the GPS may work in places where the GPRS does not work.

Automated Toll Collection System in Highways

In most of the countries, private investors carry out infrastructure development. Constructions of roads and highways are one such major infrastructure development activity. The investor constructs a highway connecting different locations and once the highway is opened up for public use, a fee is collected from every vehicle using the highway and this is called toll. When the highway is constructed and opened, vehicular traffic also increases leveraging the infrastructure availability and manual collection of toll may be a challenge both for the infrastructure companies and also users; more particularly, for the regular users of the highway. An automated toll collection system is a system whereby an electronic device is provided with a unique identification number for a very small fee, which needs to be fitted/fixed/kept in the car. The regular user of the highway opens an account with the infrastructure company's information system provider and pays the toll charges in advance, providing the unique number of the device. The toll gates will have a sensor, which is capable of capturing the unique number of the electronic device fixed in the car, even if the car is running at a very high speed, and transmit the data to the server. Then based on usage, the user account is debited with the toll charges. This is popularly known as EZ Pass in the United States of America. In case a car passes through the EZ Pass lane at high speed, without fixing the electronic device, there are high-technology cameras are capable of capturing the photograph of the moving car (even at high speed) along with its registration plate and a notice is sent to the owner of the car for payment of the toll fee for usage of the highway.

Retail Store Solutions

This is an example of an industry-specific solution for the retail industry. With this system, critical processes such as inventory management, billing, shelf-space management, etc. are carried out for a retail chain. There are several providers of retail solutions such as Toshiba (previously Toshiba Retail Solutions was owned by IBM; sold to Toshiba), SAP, etc.

Travel Management System

This is typically a software solution provided for managing the operations of a travel-related service provider. Travel-related services include air or rail booking, visa processing for international travel, hotel booking, cab booking for local transport, etc. Travel management software is capable of addressing all these needs. Typically, these systems are interfaced with system business partners of the travel management service provider. These business partners include airlines, cab-service provider, consulate offices, hotels, etc. Besides, automated communication will be sent to the customer on the status of the travel arrangements. Some of the global players in travel management services offer self-service portals for their customers to place requests for travel management needs.

Applications for Educational Institutions

Activities of educational institutions such as schools, universities, and colleges are nowadays automated through software. This application takes care of all the operations of an educational institution such as student registration and admission, fee collection, library management, hostel management, online courses, virtual classes, etc.

Document Management System

The business expansion nowadays has gone beyond the geographical boundaries of countries with a high volume of cross-border transactions. As the volume of transaction explodes, there is a strong need for managing documentation needs in an automated way. A document management system helps in managing the lifecycle of a document, through predefined configurations for a business enterprise. IBM Filenet and Documentum are some of the examples of document management systems.

Several Other Systems

There are several other systems that help the business enterprises and other institutions in managing and automating their business operations such as CORE banking system, railway reservation system, stock exchange systems, etc.

7.4 Information Processing

In everyday life, we see data being processed to meet the need of the user. The need is called information. It could be a simple transaction, where we shop for materials and pay the shopkeeper and collect the balance, if any. We quickly and mentally calculate the cost of the material we have purchased factoring the unit price of the material and the quantity purchased and pay the shopkeeper. If we collect the balance, we quickly check whether the money returned is correct, subtracting the cost of the material purchased from the money we paid to the shopkeeper and the actual money returned by the shopkeeper. We process data such as the quantity purchased, price per unit, cost of all material, money paid to the shopkeeper and the balance collected and the output amount payable to the shopkeeper and the balance payable by the shopkeeper. We can call this as manual processing or mental processing.

Information is the processed data, which is value for the user. In each transaction, there is data, which has to be processed for generating information.

Manual Information Processing Cycle

Business transactions have been happening for ages, whereas computers have come about less than a century ago and the commercial application of computers has taken place for less than 50 years. Before that, there were business, transactions, data, process and information.

Typically the information-processing cycle of manual information processing involves three stages: (1) Input of data (2) Processing of data and (3) Output of information.

In certain activities, we still carry out manual processing and not automated or computerised ones. Examples are

- ✓ Grading of answer papers of an examination, where the questions are answered with descriptive answers.
- ✓ Teaching: though the method of delivering courses has been automated such as online courses, virtual courses, etc., still, the teaching session is manual.
- ✓ Surgery performed in an operation theatre by a surgeon
- ✓ Ticket examined by ticket examiner/conductor in a train
- ✓ Billing done by a small medical shop, where there is no computerisation of inventory and the billing done
- ✓ Manual maintenance of books of accounts in small and micro business

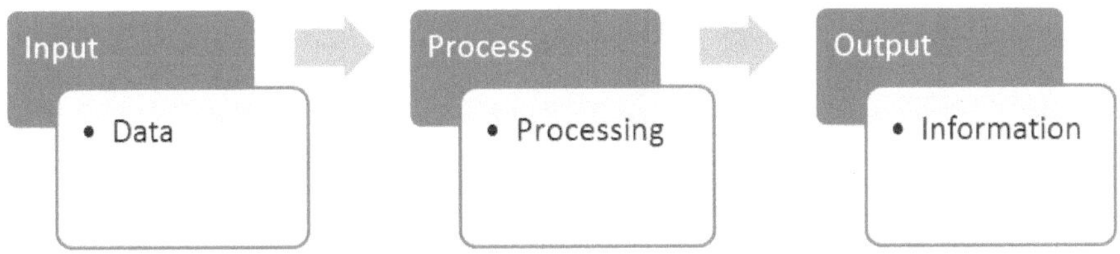

Diagram 5.2 Manual information Processing

Input: Record the data in register
Process: Summarise information
Output: Present information to the appropriate person

Computerised Information Processing Cycle

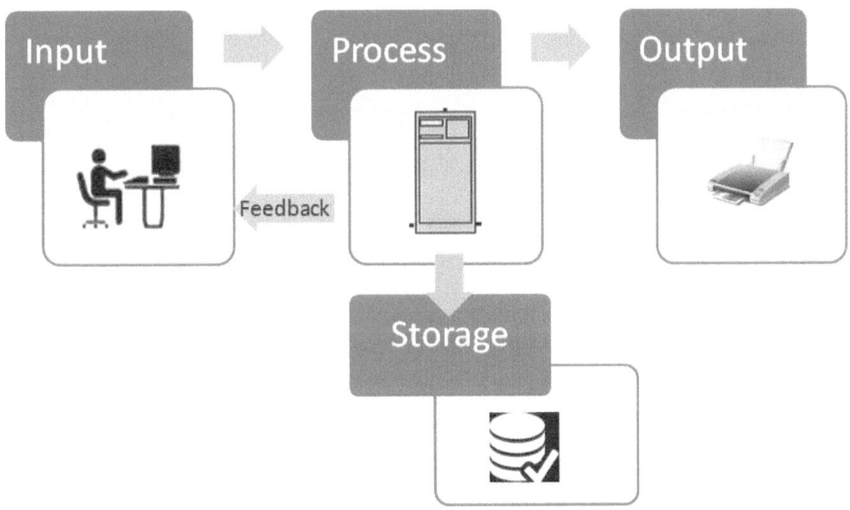

Diagram 5.3 Computerised Information Processing

Input: Entering data into computer
Processing: Data processed by computer through a program
Storage: Saving data, output for future use
Output: Presenting the results

If we look at the phases of work performance since the ages, most of the work was performed manually. The Industrial Revolution caused mechanisation of work and information and Internet technologies have caused automation. More and more business operations are getting automated with more and more new technologies being invented.

7.5 Delivery Channels

Delivery channel means the mode through which something is delivered. For example, entertainment can be delivered through a theatre, cinema, television serial or through the Internet medium. All these are delivery channels. In education, delivery channel could be classroom lectures, online classes, virtual classes and each of them is a delivery channel.

In business, a delivery channel means the delivery of product or information.

Information delivery channels in a business are

- ✓ Intranet
- ✓ Email
- ✓ Internal newsletter
- ✓ Staff briefing
- ✓ Noticeboards in a common area
- ✓ Manual or industry magazine, etc.
- ✓ Handheld devices
- ✓ Social media

Product delivery channel in a business could be:

- ✓ Delivering through a retail unit/shop
- ✓ Direct marketing
- ✓ Delivering through chain departmental stores
- ✓ Online sales and delivering to home

Importance of Delivery Channel

Delivery channel or medium is very critical for ***effective delivery*** of product/information. Delivery channel and medium selection is very important for communicating certain sensitive information or delivering certain critical products.

Planning for appropriate channel is also critical because business and customer needs will demand different channels based on the situation.

In reality, businesses adopt ***multiple channels***, using them for the delivery of products or information.

General Criteria for Choosing the Delivery Channel for Information

1. **Understand the Business and Employee Needs**: This is very important as each business has its own delivery mechanism, and also each situation will compel different delivery channels.

For example, if you have to communicate to a larger workforce in a manufacturing location, posting the communication on the intranet is unlikely to reach as most of the workers in a manufacturing plant will be working on manufacturing machines and may not have access to Internet. Here a conventional 'Town Hall' meeting or 'All-hands' meeting could be a more appropriate choice, besides displaying the communication on the noticeboard. Again, it depends upon the message to be conveyed. Say for example, if some factory workers are going to be laid off, it is a sensitive matter and needs to be communicated through an appropriate channel. In this situation, multiple channels can be chosen.

2. **Traditional Channels** are still useful and not to be discarded. For example, if training is to be delivered to the employees as part of organisational development, it is good to have a classroom session, where people will spend time away from work, will have the opportunity to network with other employees; a classroom session is always more interactive than other modes. Similarly, enterprises still have 'Town Hall' meetings and 'All-hands' meetings as part of the mass communication strategy to provide updates about the overall business.
3. **Internet and Intranet** are really good channels. But, still these channels have their own limitations. The very first risk is that the targeted audience may not read the communication. Sometimes, this internal information may be circulated outside by just copying screenshot of the message; it may then go to an undesirable audience. Therefore, the enterprises are careful in choosing this channel. Often, the enterprise chooses Internet or intranet only for selected communication and still prefers to use conventional delivery channels for various reasons.

General Criteria for Choosing the Delivery Channel for Products

In general, products are delivered through both conventional and modern channels. The main reason is that the customer needs are different. Certain customers may prefer a personal shopping experience and others may prefer an online shopping combined with a delivery to home.

Conventional

Typically, conventional delivery channels are in the form of the brick-and-mortar model. For example, delivering through a **retail** outlet.

Certain businesses take the approach of **direct selling** to customers, where the sales personnel will visit homes of the customers and promote products by performing demonstration.

Modern

Internet triggered the practice of **e-Commerce and online sales** making them very popular. The critical delivery capability for an e-Commerce set-up is to invest in a backend logistics delivery channel network.

Tele-marketing sales with home delivery set-up is another channel. Tele-marketing can be either the business approaching customers over phone or it can be the promotion of products with exclusive 'Teleshopping' shows through television. In the teleshopping case, the customer has to reach out to the business over phone for placing an order, after getting information about the product and business contact details through television.

Business combines both the traditional and modern channels for promoting products and business.

7.6 BPA Controls

Establishing necessary controls in an automated environment is a basic need. It becomes even more important from an accounting and audit perspective. Establishing an internal control system is fundamental of any business control. A business environment that has been automated with the help of information and Internet technologies is not an exception.

Control Objectives

The objectives of BPA controls are as follows:

Authorisation

Every transaction in the business is carried out in line with the established processes and procedures and there is the authorisation of the competent authority in the business organisation to carry out the transaction.

Completeness

This is to ensure that all business transactions are included in the business systems and considered in accounting and that also even the inadvertently wrong transactions have not been considered.

Accuracy

This is to ensure that all recorded transactions are accurate within the given business context. This is very important in an automated environment, where external data flows into the system and becomes part of the system transactions.

Validity

This is to ensure that all transactions are within the business context based on the economic activities of the business. This goes along with the authorisation criterion.

Physical Security

All the real and information assets have to be safeguarded with physical security. This is critical even in an automated environment. There are various methods by which technical security can be offered to a business environment, besides security to the business facilities. Based on the importance of this control criterion, in most of the ERP when a transaction is carried out, one cannot undo the transaction. If the effect of an erroneous transaction is to be nullified, one has to pass and post an accounting journal document for correcting the error.

Error and Exception Handling

These criteria call for a system to have a validation-control procedure to handle errors and exception. Most of the ERPs will have exhaustive validation procedures to ensure that before posting a document, all validation parameters have been verified and the data provided is accurate.

Segregation of Duties

This control criterion was derived from the conventional internal control system. This is applicable for an automated environment also. Segregation of duties is one of the verifications carried out in an information system audit.

BPA Control Classifications

Business Process Automation envisages the following controls:

1. Application control
2. Internal control

Application control

Application control refers to the controls that have to be established in a system. When automation happens, unauthorised users have to be prevented from accessing the system through the possibility of unintended errors in the system that might have crept into the system or omissions, which can adversely impact the transactions and consequently impact the accounting records and eventually financial statements.

The controls explicitly and systematically set up in the applications and the environment are called application controls. More details of the application controls will be discussed later.

Internal control

Broadly, internal control refers to a control mechanism that has been set up in an enterprise to minimise risks, if not avoiding risks all together. An internal control system helps in preventing and detecting frauds in the organisation and business enterprises. Accurate financial reporting relay is based extensively on internal controls.

The statement on accounting 315 (SA-315) deals with the auditor's responsibility to obtain an understanding of the entity and its environment and using that understanding, identify and assess the risks of material misstatement at the financial-statement level and assertion level.

An internal system is said to be well designed and in control when the system facilitates

- ✓ That all transactions are carried out in accordance with the policies and procedures laid down by the organisation
- ✓ That all transactions are duly approved by the competent authority
- ✓ That resources of the enterprises are channellised in line with the organisational goals
- ✓ That all the assets of the organisation are safeguarded

In an enterprise, controls are established at different stages and the nature of control is based on the stage at which the control is set up. The nature of control is classified as follows:

- **Preventive**: These controls prevent mishaps in the organisation and loss of assets. For example, installation of firewall prevents the loss of information assets.
- **Detective**: Detective controls check whether any errors have crept in. For example, a validation check in the program helps to detect the errors in the data input.
- **Corrective:** Corrective control is to fix the error identified. In the olden days of offline data entry, where no validation check was possible, before running the inputs for processing there was a step called edit check where all the data entered was printed out and thoroughly checked for errors.

In certain contexts, corrective control also means the steps taken by the organisation in minimising the loss, if not its avoidance. For example, taking a forward cover for minimising or avoiding loss against fluctuating foreign exchange.

Various Application Controls

Broadly the applications controls are classified as below:

1. Access control (also known as boundary control)
2. Input controls
3. Process controls
4. Database controls
5. Output controls

1. Access Control (or Boundary Control)

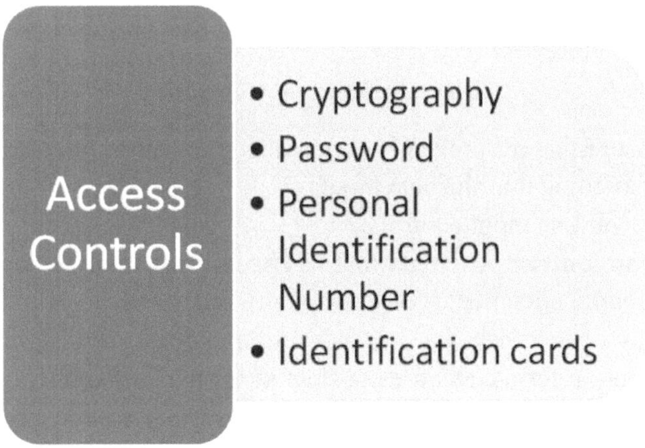

Diagram 5.4 Access Control

Different types of access controls are discussed below:

- **Cryptography** is a method by which the data/messages are encrypted when transmitted and decrypted by the recipient after the receipt to read the text messaged.
- **Password** is the mechanism for obtaining the credential validation of an authorised user. Once a password is entered correctly, the system assumes that an authorised user is accessing the system.
- **Personal Identification Number (PIN)** is something similar to a password, but it is limited in length and generally will contain only numeric.
- **Identification Cards** are shown to the user to enable the user to pick up the right identification card that was set selected and set up by the authorised user, while setting up the access. If the correct identification card is selected, the system believes that the user accessing the resources is an authorised user.

2. ***Input Controls***

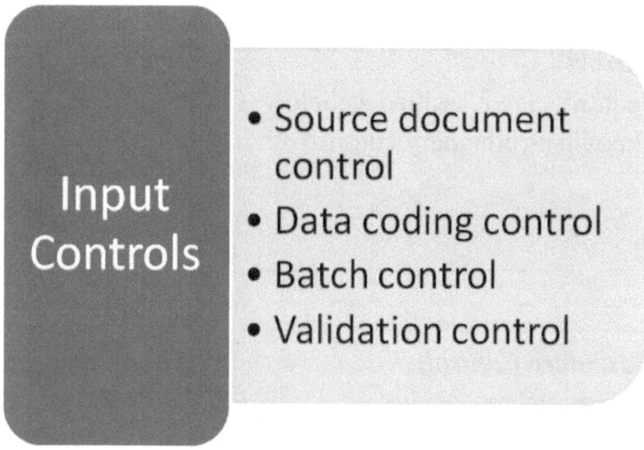

Diagram 5.5 Input Controls

Input controls ensure that all preventive approaches are taken against controlling even the inadvertent errors from going into the data input.

Following are the various input controls:

- **Source Document control:** Any data that has been input into the business system should be backed by a source document. For example, to recognise revenue and to ship material, the business needs a sale order. Sale order is a source document. Similarly, purchase order, income-tax declaration forms, etc. are considered as source document. It is important to be cognisant about the validity of a source document, as it could be forged and produced.
- **Data Coding control** reduces the errors while feeding the data. This was very prominent when the offline data entry was in practice. There were different types of errors committed during the offline data-entry operations. Even now the types of errors listed below are valid.
  - **Addition** of an extra character in a code—for example, 103 may be fed as 1103.

- **Truncation** is where a character or a number is missed—for example, 103 may be fed as 03.
- **Transcription** is an error in which one of the characters or numbers has been wrongly entered—for example, 103 may be fed as 703.
- **Transposition** is an error, where the position of letters or numbers gets interchanged—for example, 103 may be fed as 130.
- **Double transposition** is an error where one or more character separates the number of transposed numbers—for example, 103 might be fed as 301.

Errors in feeding the data have serious consequences. Hence, they need to be corrected and corrections have to tracked. For this purpose, a control called check digit will be set up—a number will be generated by the computer applying certain mathematical operations on one of the data, say serial number of the data, and will be added to the data used for check digit generation. For example if a serial of a data is 100, a check digit will be added at the end after performing certain mathematical operations on the data. In this example, the check digit can be the multiplying of each number with its position.100-1 is 100th position, 0 is in 10th position and another 0 is in the unit position. Mathematical operation will be (1X100)+(0X10)+(0X10) = 100 and the addition of these digits will be done to form a single digit, in this case 1. The check digit in this case is 1. The serial number will be entered as 100-1; numbers after hyphen being the check digit.

- **Batch control** was prevalent during the offline data entry days as batch processing was happening. Even today, certain batch processing operations are controlled through this batch control. Typically, a batch number along with the time and date of processing the batch will be forming a part of batch control for identification. Along with the batch control identification, additional information on the batch will be recorded for control such as number of records, total monetary value of the data processed, etc.
- **Validation control** is a preventive approach to an erroneous data feeding. The control is established when the data is entered or captured. In other words, validation control is close to sourcing the valid data. As we are aware, a record will have fields and an error can be at the field level and if so, then the record is defective. In addition, a record-level error could occur, where a field itself is completely missed to be fed in with data. There are two levels of interrogations to control the data error as follows:
    - **Field-Level Interrogation** is of the following types:
        - **Limit check**, where the length of the field is limited by characters
        - **Picture check** is where only numerals and not characters are allowed and if any character is fed, the system will caution the user as error in the data.
        - **Valid-Code check** is where each field will be having a valid code and the data entered will be compared with the predefined and available valid code to filter any error.

- **Check Digits:** A data entered in a field along with the check digit will be verified by the data-entry system to check whether the data entered is correct by generating the check digit and comparing the same with the check digit fed by the user at the time of data entry.
- **Arithmetic check** is carried out, if it is possible to establish a relationship between the data. For example, if the data fed in is the quantity, price and total value of an item, multiplying quantity and price and comparing with the data fed as the value of an item can perform the arithmetical check.
- **Cross-checks** is the process, where the data entered is checked with the similar data in different files. In the above example, the price keyed in can be checked with the price already recorded in the price file with the corresponding item.

- **Record Level Interrogation** is of the following types:
  - **Sequence check** is to check whether all the data entered is in an appropriate sequence.
  - **Format Completeness check** is where all the fields have been verified to see that no field is left blank and that all data is in an appropriate format in the respective fields.
  - **Redundant Data check** verifies whether the same record is repeating more than once.
  - **Password check:** Every record entered should be suffixed with the user's entered password.

3. Process Control

Controls established while processing to verify and cross-check that the data, processing and the results of the processing are in line with that has been planned are called process control.

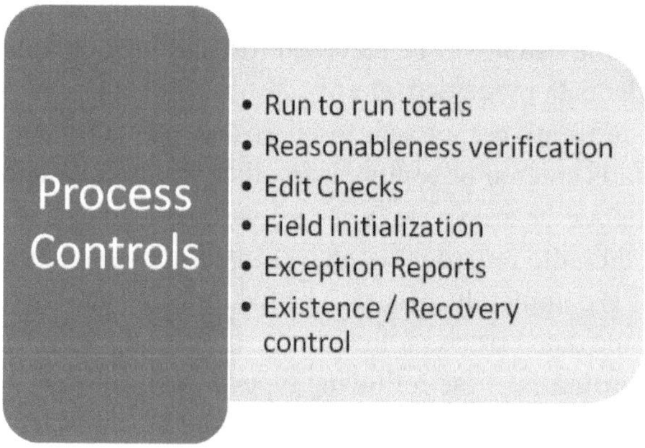

Diagram 5.6 Process Controls

- **Run-to-run Totals** are a control that is established for verifying whether the processing is being carried out in order. Each run totals and cumulative totals from the start of the period (say a financial year) are provided; these can be cross-checked with the previous run and current run and update all the processing. These are typical controls of batch processing.

- **Reasonableness verification** is cross-verifying the available data for checking that the overall information has been processed. For example, in a CORE banking system, it could be checking the interest paid on a deposit based on the percentage with the amount of deposit and confirming that some randomly selected sample records prove to be correct.
- **Edit checks** is the same as discussed above. It is the corrective classification of control, where the offline data entered in a medium is printed out and verified for the correctness of all data. These processes were prevalent during the mainframe computer days for batch processing.
- **Field Initialisation** is a process where before every run of the program, all the fields are made zero to ensure that the processed data for the batch is started afresh. Cumulative values will be collected from a separate file for computing the batch-control run totals. Field initialisation also helps technically in data overflow avoidance.
- **Exception Reports** are generated for providing the data not considered for processing. Typically, in the programs certain criteria for data validation will be given for consideration of processing and certain data not meeting the criteria will be segregated and reported with reasons. These are called exception reports.
- **Existence/Recovery controls** are precautionary steps to recover the data in case of failure of the system while the processing the data. These are the temporarily saved and backed-up data for recovery after a system failure.

4. Database Controls

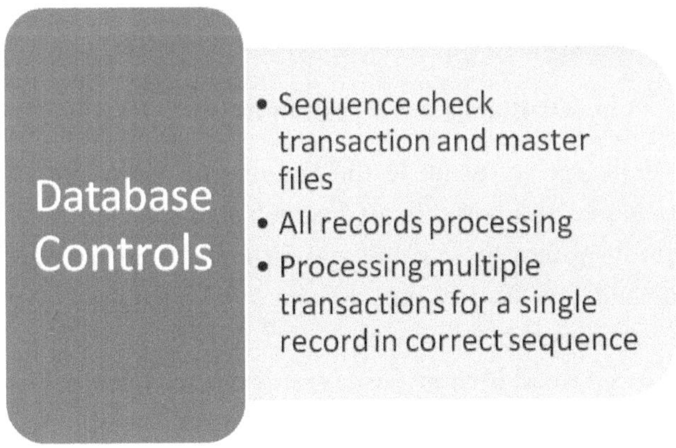

Diagram 5.7 Database Controls

While handling the data in the transaction database and/or master database, the program has to handle it sequentially and exhaust processing all the records and reckon with the end-of-file confirmation; this activity is called database control. These were needed, when the data was stored in text formats in the olden days of computers, before the invention of RDBMS. But, these controls are still valid in today. Various database controls are

- **Sequence Check Transaction and Master Files:** In this control, all the records are sequentially checked both in the transaction and master files to ensure that no records are missed for processing. Every record is considered and printed either in the main output or exceptional output.
- **All-Records Processing** control ensures that all records are considered till the system encounters the end-of-file (EOF) status. This is again to ensure that all the data is considered.
- **Processing Multiple Transaction for a Single Data** is common for a master data where there can be multiple child transaction data. All child data have to sorted and sequentially processed which assures a control while processing the multiple transactions for a single data.

5. ***Output Control***

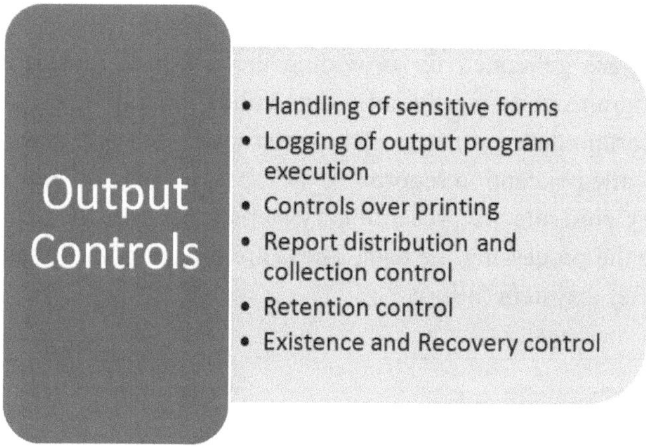

Diagram 5.8 Output Controls

Outputs are generally delivered for usage to the end-users. The delivery medium can be in the form a printed report or saved in a portable media such as CD ROM or DVD or Flash drive or could be saved in the built-in hard disk of the system. While delivering the output, certain controls have to be established for security and these are called output controls. The various types of output controls are discussed below:

- **Handling of Sensitive or Critical forms:** We have experienced that banks issuing a demand draft or bankers' cheque or a company issuing dividend warrants in a pre-printed stationery cheque that are set up in the printer to print cheques, warrant etc. are sensitive forms and there is a strong need for tracking the usage of these forms to avoid their misuse. Generally, a log is created for the usage of forms, their cancellation and safe destruction.
- **Logging of Output Program Execution** is critical when the critical output programs are executed. For example, if an interest calculation program is run in a banking system once in every 180 days, it has to be logged. Even an inadvertent run of the program more than once will result in crediting interest twice for the same period.

- **Controls over Printing** are another important aspect of output control. This control is set up in the program itself, enabling only the authorised users to print the information or data. This is exercised to ensure that confidential information is going into the hands of only the right and authorised users.
- **Report Distribution Control** is critical because, in the olden days, line printers were centralised and reports were printed by the MIS or IT department and distributed only to the authorised users. The tracking of the distribution to authorised users was assured through this control. This control still holds good and valid. In the current online real-time scenario, authorisation to run reports and view the results on the screen is restricted through role-based authorisation.
- **Retention Control** is about the retention of output records and their systematic destruction. The policy of the organisation and regulatory needs have to be taken care of and accordingly, this control is established.
- **Existence/Recovery Control** are needed in two scenarios. Firstly, when the output is being generated and at that time, if the system fails, the output should be safely residing in the system for reprinting. Secondly, after printing, if additional copies are needed or the originally printed records are destroyed, reprinting will be required from the saved report. Typically, reports are stored in the reports spool in mainframe computers. That assures their existence and the recovery control.

Section 8
Access and Identity Management

8.1 Background

Safeguarding the information system, data and information from unauthorised users is of paramount importance for every business enterprise. The main reason for this is that every business is attempting to differentiate itself from the rest of the players in the industry. From that perspective, if any information about the enterprise's competitive advantage is leaked, it will be detrimental to it.

The access and privilege control functionality in a business system takes care of this and tries to prevent unauthorised users from accessing the IT resources of the enterprise.

There are several technical methods through which the user access is managed and these are (1) Identity Management (2) Authentication (3) Authorisation and (4) Accountability.

8.2 Identity Management

This is a two- or three-step process through which the identity of an authorised user is verified. This could be, for example, through the RSA Secure ID. "Rivest Shamir Aldeman" is the full form of RSA. In this case, the authorised user is provided with an RSA token, which generates certain random numbers in a fixed frequency—say, 60 seconds. When a user wants to log on to the enterprise IT environment to which the user has been authorised and provided with an RSA token, the user has to log on to the system with the user ID along with the RSA token-generated latest random number. The system environment verifies the details with the enterprise server, where the similar RSA random number is generated and on confirmation of the authorisation, the user is allowed to log on to the IT environment. The RSA token diagram is provided below:

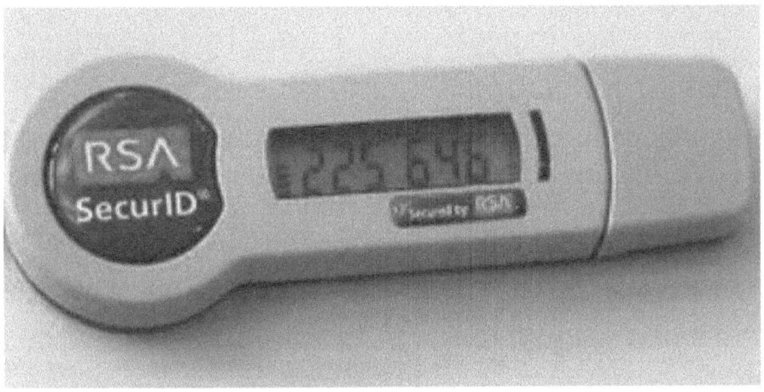

Diagram 8.1 RSA Secure ID Token

The RSA token ID is in addition to the user-identification code and password.

In certain systems, an image is expected to be chosen by the user at the time of registration and every time when the user is logging on, the appropriate chosen image has to be selected from the several images shown to the user at the time of login, to identify the user genuineness.

An example of image identification is provided in the diagram below:

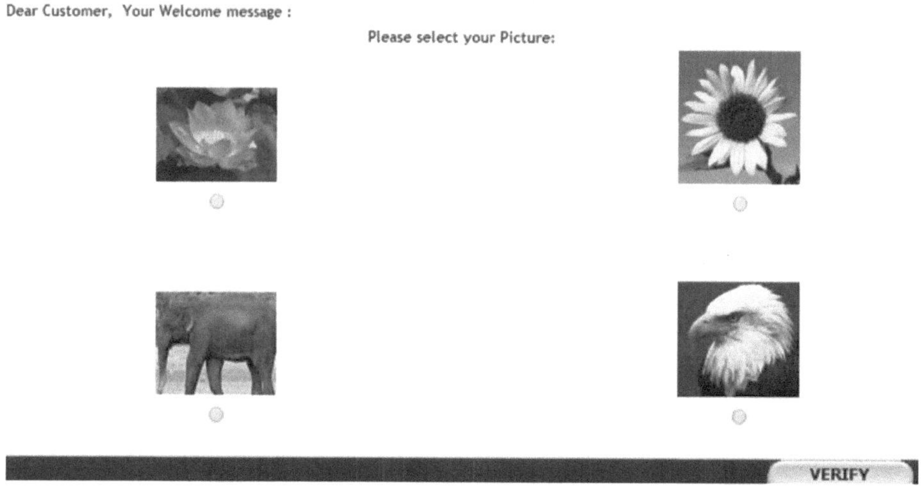

Diagram 8.2 Image-Based Identity Management

Certain transaction systems are expected to be used by human beings only and in order to prevent an automatic system or robots accessing those systems, a CAPTCHA test is carried out (CAPTCHA—**C**ompletely **A**utomated **P**ublic **T**uring test to tell **C**omputers and **H**umans **A**part). In this case, an image showing numerals or alphabetical characters is shown to the user and the user is expected to record the shown numbers or characters in the appropriate section of the scheme. This step is possible only by the human user and in that way the users are differentiated between a human user and a non-human user. An example of the CAPTCHA identity verification is given in the diagram below:

Diagram 8.3 CAPTCHA-using Identity Management

8.3 Authentication

Authentication is the process of identity confirmation. When the user is able to successfully identify oneself, the system authenticates the identified authorised user to access the IT resources of the given IT environment, subject to authorisation.

8.4 Authorisation

Authorisation is the process by which the user is allowed to perform certain activities or transactions that he/she is authorised to perform. Authorisation basically controls all the accesses that an authorised user possesses.

Another context of authorisation is the approval from a competent authority in the enterprise to set up a new user as an authorised user. While setting up a new user, the authorisation approval will contain what are the activities that the user needs to perform as part of his/her official responsibilities, for which only is the authorization to be provided. The authorisation process is generally practised with a concept called 'Enforce least access or least privilege'. This means, no access is provided unless the need is proved.

There are two kinds of authorisations:

Role-Based Authorisation/Role-Based Access Control (RBAC)

In order to avoid any kind of discretionary decisions on providing or revoking authorisations, the enterprise can evolve a policy for role-based authorisation. In this case, the roles of the different organisational personnel are defined and corresponding access to different system resources are assigned based on the need to perform their activities.

Rule-Based Authorization/Rule-Based Access Control (RBC)

This control is based on the concept of internal control established within the enterprise. For example, a person may have authorisation to approve the expense claim of his entire staff based on his role as manager. However, a rule can be set that no one can approve one's own request for expense reimbursement.

8.5 Accountability

The computer system has the capability of keeping track of the users' activities. However, it is optional for the user enterprise to switch on this feature and it is generally recommended to switch it on for various purposes including audit.

The accountability comes in two ways:

*1. **Enterprise Responsibility***

The enterprise needs to set up access control defined for each user and to enforce it. User validation is to be periodically done and reconciled. Also, the enterprise should periodically verify the activities performed by the users through the log created in the system and provide necessary education to the users as part of the IT training about their responsibilities.

*2. **User Responsibility***

The user is expected to adhere to the enterprise's IT policies and ensure that the IT resources of the enterprise are properly used in accordance with the IT policy. Based on the education and IT training provided by the enterprise, the user should strictly follow the data security and privacy processes laid down by the enterprise. In event of doubt or upon observing certain suspicious activities, the matter should be reported by the user to the designated appropriate authority in the enterprise.

Principle of Least Privilege

The principle of least privilege is based on the concept of conservatism in providing access to anyone. In any enterprise, the user access is managed through standard templates for various roles and based on the role assigned and approved, the appropriate template role will be copied and access will be provided to the new user. One such standard role definition is 'least privilege user'. In case of a 'least privilege user', the user will be able to access the enterprise's IT resources and go around different pages on the Intranet and perform basic functions such as submit timesheet or expense report, but will not be able to perform any other computing activities. If no role has been defined for a new employee, by default the employee will get the profile of 'least privilege user'.

8.6 Payment Mechanisms

As the Internet commercial usage caused creative business modelling, so was with the payment methods. Broadly the transactions can be (1) between customer and an online merchant, who offers goods or services and online payment processes using credit cards or debit cards and (2) between customer and bank—like online banking or mobile banking or ATM transaction as against traditional branch banking. This also iincludes Point-of-Sale (PoS) transactions.

8.7 Payment Process of an Online e-Commerce Transaction

The diagram provided below explains the full cycle of an online e-Commerce transaction from a customer's perspective:

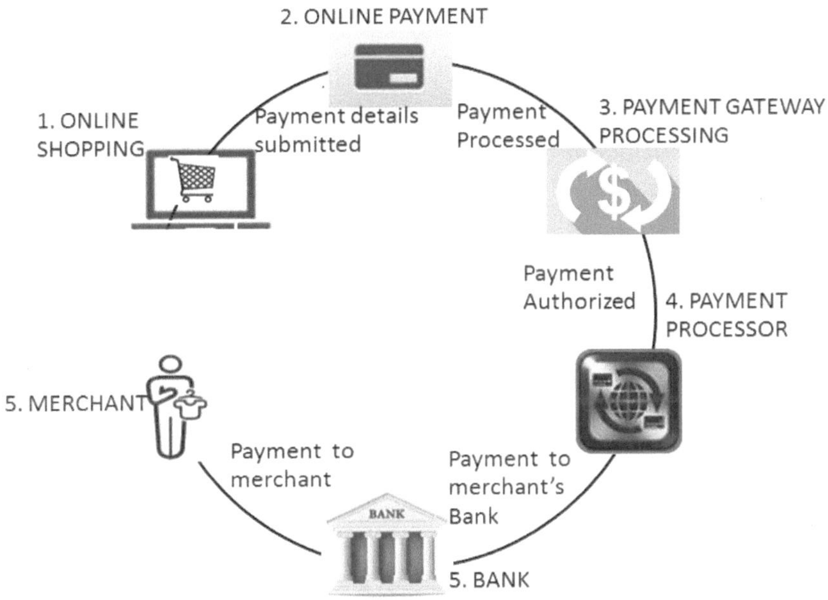

Diagram 8.4 Online e-Commerce Transaction from a Customer's Perspective

As the customer places an order and provides the payment transaction credentials, the merchant site or payment site connects to the payment gateway and processes the payment approval with the customer bank; once the approval is received, the response on payment confirmation is sent to the customer and money is transferred to the merchant's account.

8.8 How does a Credit Card Transaction Work?

Step 1: The bank issues and activates a credit card to a customer on his/her application request and after approval.

Step 2: The customer presents the credit card information to the merchant site or to the merchant from whom he/she wants to purchase a product/service.

Step 3: The merchant validates the customer's identity by asking for approval from the card brand company.

Step 4: The card brand company authenticates the credit card and pays the transaction by credit. The merchant keeps the sales charge slip.

Step 5: The merchant submits all the credit card sale charge slips to the acquirer banks for payment to him/her. This is called batching process which is typically done at the end of each day.

Step 6: The acquirer bank requests the card brand company to clear the credit amount and gets the payment net of service charges of the card brand company. The acquirer bank pays the merchant after recovering its service charges.

Step 6: Now, the card brand company asks the issuer bank to clear the amount and the amount gets transferred to the card brand company.

8.9 Merchant's Claim Clearing and Settlement

The diagram given below is self explanatory as to how the clearing and settlement of claim happens and how the card-holder account gets debited.

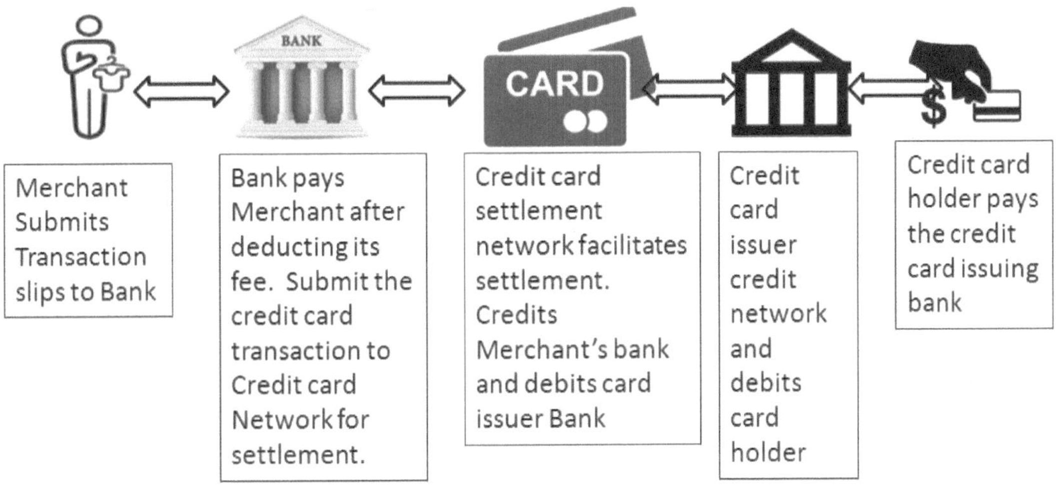

Diagram 8.5 The Clearing Settlement Process

8.10 Transactions between Customer and Banks

Point-of-Sale (PoS) and ATM Transactions

The customer swipes a credit card or debit card (also called ATM card) for ATM transaction or at PoS. If it is a credit card, the customer credit-card account is debited and if it is an ATM or debit card, the customer checking or savings account is debited with the transaction.

Credit-card transaction is the same as that of an online e-Commerce transaction explained above.

In case of a debit card or an ATM card, whether transacted at an ATM machine or PoS location, the ATM system or merchant IT system communicates with the issuing bank's CORE banking system and seeks approval. The CORE banking system of the issuing bank verifies the customer account and balance. After debiting the customer account, the approval is sent back to the ATM or merchant system at PoS. Then, the ATM system disburses cash to the customer or the merchant ships or hands over the products to the customer or extends service, as the case may be.

Smartcards

Smartcards have been recently introduced, replacing the magnetic-stripe cards, for ensuring better security.

Earlier in the magnetic-stripe cards, a magnetic strip on the backside of the card contained all the details about the card, which was transmitted for authentication and approval, when swiped.

In a smartcard, instead of a magnetic stripe, a chip is embedded and the chip contains the details of the card.

A smart card can be a contact card or Contactless card.

Contact card

In this case, the smartcard is inserted into the device of the merchant or the ATM machine and the merchant device or ATM machine reads the data from the smartcard.

Contactless card

The smartcard is waved near the merchant device and the data is transferred to the merchant device through the NFC (Near Field Communication) technology.

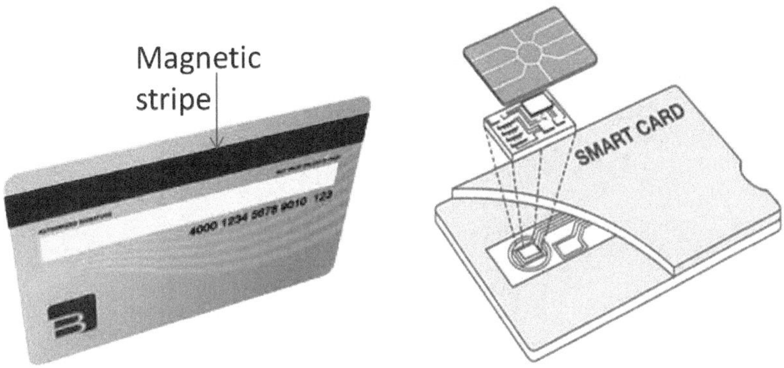

Diagram 8.6 Magnetic-Stripe Card and Smartcard

Electronic Wallet

An Electronic Wallet (also called as a digital wallet) refers to an electronic device that allows an individual to make e-Commerce transactions. This can include purchasing items online with a computer or using a smartphone to purchase something at a store. Increasingly, digital wallets are being made not just for basic financial transactions but to also for authenticating the holder's credentials. For example, an Electronic Wallet on a phone could potentially verify the phone number of the device to verify with the registered phone number of the customer. The components of Electronic Wallet are that of the "digital wallet" not as a singular technology but as three major parts: "the system"—includes the entire electronic infrastructure; "the application"—the software that operates on infrastructure and "the device" that is used by the individual.

Section 9
Advanced and Emerging Technologies

9.1 Advanced Technologies

The emerging technology includes wireless technologies. There are two important wireless technologies that are in extensive usage as given below:

Bluetooth

Bluetooth is a technology for wireless data communication within the radius of 50 metres or 164 feet.

Using Bluetooth technology, the following devices can be connected wirelessly:

- Keyboards
- Mice
- Printers
- Handheld devices
- Laptops/desktops
- Digital cameras
- Remote control devices

Wi-Fi

Wi-Fi is another wireless technology through which a system can connect to a Wi-Fi router with the radius of 32 metres (120 feet). Wi-Fi is based on conformance to IEEE 802.11 standards, which use radio technology that is considered safe and fast.

Large corporations are offering Wi-Fi connections in their premises. Nowadays, homes have a Wi-Fi facility to connect to the Internet.

As a part of governance, some of the smart cities are offering public Wi-Fi connectivity in all parts of the city.

Other products under relatively newer technologies are

Laptop/Notebook Computers

A laptop is a mobile computer weighing about 3 kg, which is capable of performing all the functions of a personal computer. A laptop is fixed with a flat-panel technology monitor and an in-built hard disk. Certain types of laptops are smaller than the others sand these are popularly known as notebooks.

Tablet PC (such as Apple iPad)

A tablet PC, simply called as tablet, is a general-purpose personal computer in a single panel with touch interface. Touchscreen is the unique characteristic of a tablet, where the touchscreen itself is an input device.

Some of the features of a tablet are

- ✓ Input method is through touchscreen
- ✓ Size is smaller than a laptop, which can be held in hand
- ✓ Battery life is longer as the design of the tablet is with low power requirements
- ✓ Tablets are capable of storing data generally between 16 and 64 GB
- ✓ Performance is low as the processors are an enhancement of the smartphone processors; therefore, they can be used for handling mails and browsing internet and making notes and using for some form of entertainment. They are not suitable for running major applications. However, like any other mobile device, a tablet can be used for accessing enterprise resources through mobile access.
- ✓ Software: Tablets are produced either with iOS or Android. iOS are used for iPad by Apple and Android is used by the rest of the tablet brand owners.
- ✓ A tablet can connect through wireless through Wi-Fi and Bluetooth.

Smartphone

A smartphone is with an in-built mobile operating system capable of performing communication functions, computing functions and providing entertainment media for the user. A smartphone can be defined as a traditional PDA with the functionality of a cellular phone.

New models of smartphones are with high storage capacity and with high resolution touchscreens.

The leading operating systems on which smartphones are working are iOS of Apple, Android of Google and Blackberry OS of Research in Motion.

iPOD

It is a modern device on which media is played. In a nutshell, it is a modern media player. iPod is the product of Apple released in 2001 with Macintosh-compatibility iTunes. Recent redesigned versions include iPod Shuffle, iPod Compact, iPod Nano and iPod Touch with touchscreen user interface. The recent versions are not only Mac-compatibility iTunes, but also Windows compatible.

Ultra-Mobile PC

It is a very small version personal computer with a TFT display size of 12.7 to 17.8 cm diagonally. The input interface is touchscreen or by using stylus-like tablets.

iOS, Android

iOS and Android are the leading smartphone (and tablet) operating systems. iOS is the property of Apple Computers and Android is owned by Google and available for smartphone brand owners for installation in their phones for a fee. iOS is used by Apple in their product range—iPhone, iPad, iPod and iWatch.

HTC, Samsung, LG, Motorola, Sony and other leading smartphone brand owners use Android. Since Android is open source, any new player in the mobile phone industry can choose to create a device on this platform.

9.2 Emerging Technologies: Cloud, Mobile, Virtualisation, Big Data, Artificial Intelligence, Wearable and IoT

Cloud Computing

Owning space in the IT infrastructure, platforms, applications provided by a service provider and accessing those infrastructures, platforms, applications through internet is called cloud computing.

i. Cloud-Computing Environment

There are multiple types of cloud computing environment. Broadly, these types can be classified into four categories:

- Private clouds
- Public clouds
- Hybrid clouds
- Community clouds

Private clouds: These computing environments exist within the premises of enterprises. These infrastructures are for the exclusive use of the enterprises owning them and the access is accordingly provided. IT departments of the respective enterprises generally set up these infrastructures. This helps an enterprise to have grid and virtualisation. Private cloud is scalable.

Public clouds: The public clouds are made available to general public or group of enterprises by the service provider. Third parties administer these clouds over the Internet. The service model is 'pay-per-use' model. The public cloud is beneficial for setting up the development and testing environment of the IT environment; these infrastructures are available at an affordable price and are scalable. Above all, there is no need for the incurrence of capital expenditure by an enterprise.

Community clouds: Under this model, the cloud infrastructure is shared by two or more entities of the same community. For example, government departments may share the same cloud infrastructure and services. The risk is that the data of enterprises that are mutually competing may be stored in the same IT environment as the enterprises may be part of the same community.

Hybrid clouds: This type of cloud is a combination of private, public and community cloud services availed by a single enterprise. These clouds used by a single enterprise are entitled with a unique identity and are bound together by standard data and portability of data between the clouds.

Enterprises leverage the hybrid cloud model by hosting noncore applications in the public cloud and core applications in the private cloud.

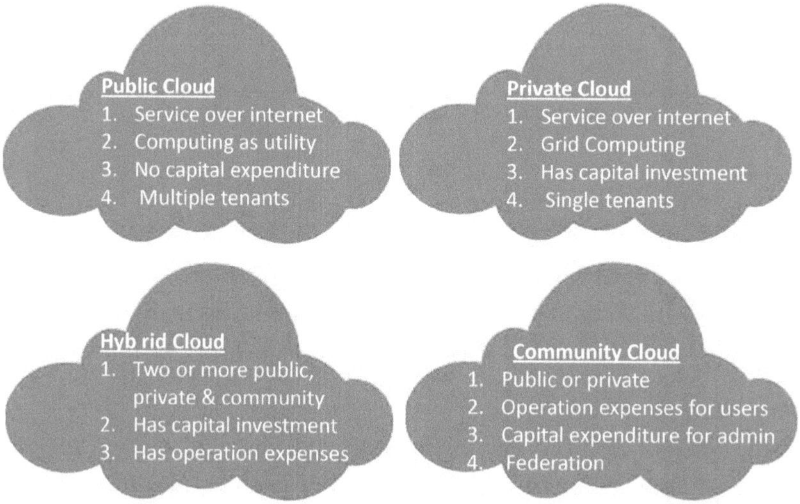

Diagram 2.3 Various Cloud Computing Environments

ii. Cloud Computing Architectural Needs

Cloud computing typically consists of two platforms: Front-end Platforms and back-end platforms. Other components of cloud computing is network which, invariably, is the Internet. All these three together form the cloud-computing architecture.

Additionally, a cloud architecture communicates with other components such as middleware, services and hosted software.

Front-End: The front-end of the cloud consists of a client's devices or a network (of an enterprise). Some cloud services may provide a standard browser-based access in which case standard web browsers such as Firefox, Microsoft Internet Explorer or Apple Safari can be used. Certain providers provide certain proprietary application for a secured interface and in such cases such proprietary applications deployed on the client device are also part of front-end components.

Back-End: Depending upon the range of services offered by a service provider, the back-end could be just a set of infrastructure with servers, storage and network or it could also include select operating systems and databases or it could include applications also. Often, back-end also includes middleware that connects networks and facilitates a transfer of data between systems.

Storage is also part of back-end.

iii. Service Models of Cloud Computing

There are several service models that are being continuously developed by providers and the popular ones are discussed below:

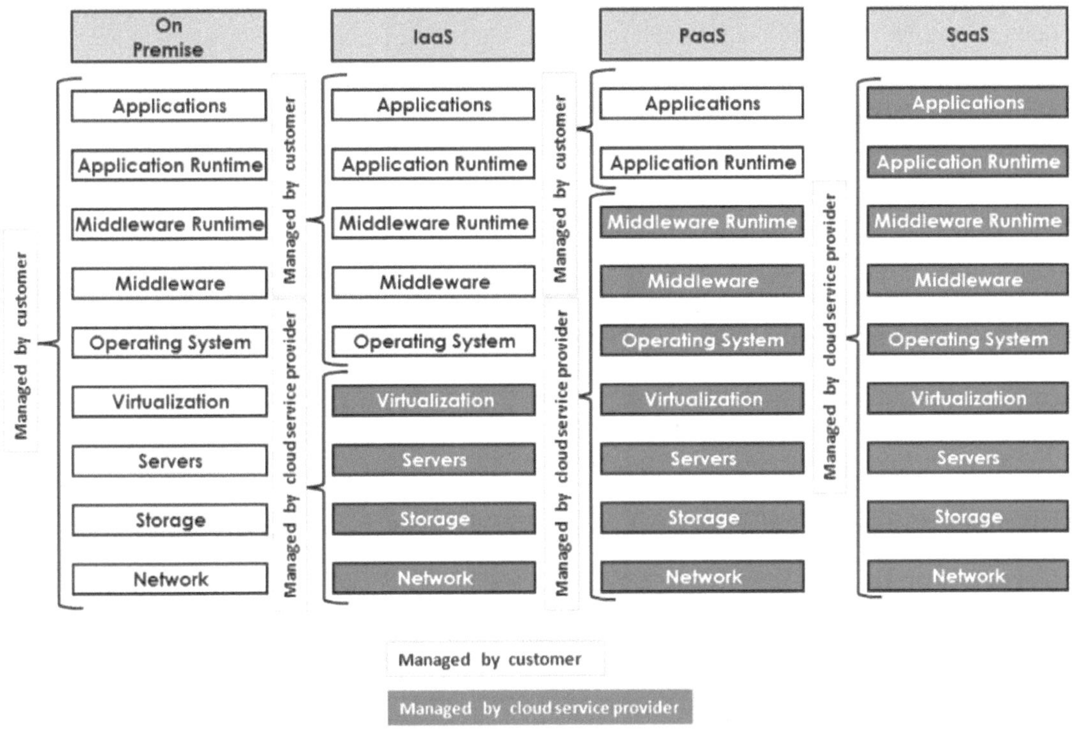

Infrastructure as a Service (IaaS)

This is perhaps the oldest model of cloud computing, where the provider offers IT infrastructure to customers as a service. The infrastructure offered as service includes servers, storage, bandwidth and other fundamental computing resources. The fee is determined based on the usage of infrastructure. It is scalable. The customers, invariably, may not know where the computing infrastructure is located.

Platform as a Service (PaaS)

This service provides a basic operating-system software along with infrastructure. This helps the enterprises to acquire a ready-to-use platform for installing development software and to move on with the development activities. Certain providers offer databases and payment services along with platforms as a service. The only disadvantage of PaaS is that there is a possibility that the providers may not guarantee a fixed-platform environment, where they have set up a development environment.

Software as a Service (SaaS)

Under this offering, the provider offers the complete software offering on the cloud. This means that the complete infrastructure, specified platform and specified software are offered as a bundle.

This is a well-established model and most of the software brand owners are moving into this model as against the traditional licensing model. Under this model, the customers are charged with periodic subscriptions as against the license fee in the traditional licensing model. The single biggest advantage of this model is that the customer is freed from the periodic patch update and upgrade activities, which are part of the traditional license-based software ownership.

In addition to the above service models, there are some more computing facilities such as ***Networking*** offered ***as a Service***, which includes the Virtual Private Network (VPN) and mobile network.

As IT and communication converge, telephony services are fast shifting to VoIP. Based on this trend, a ***Communication as a Service*** model is also emerging. This includes the instant messaging services.

Characteristics of Cloud Computing

Scalability

Scalability is the one of the important characteristics of cloud computing. Almost on demand, a customer can demand from the provider, the additional requirement of cloud services. Cloud computing service is designed in such a way.

Pay-per-use

The service provider gets paid by the customer for usage only. This characteristic helps the customer to the highest level of benefit.

On Demand

Cloud services can be availed by a customer 'on-demand'. On a short notice, the computing environment of the choice of customers is made available by the provider.

Resilience

The cloud services are resilient in the sense that unlike in a private enterprise-owned data centre, where failure can impact upon the work, in cloud services, the provider handles the failure suitably in such a way that there is no impact on the customer service. It is possible that the provider can migrate the data and computing facility to a different environment, even without the knowledge of the customer.

Multi-Customer/Multi-Tenant

In a public cloud, various customers will use the environment. Therefore, unlike a private data centre, a public cloud-computing environment is a public environment for multi-customers. This characteristic is also called as multi-tenancy or multi-occupants.

Workload Distribution

This is a very important characteristic of cloud computing. As the cloud service provider underwrites service levels to various customers, the provider ensures the underwritten service level by an optimal workload distribution. This also goes with the resilience characteristic.

Benefits of Cloud Computing

An enterprise that plans efficiently and leverages cloud services can derive enormous benefits from cloud computing. The general benefits are as below:

No Capital Investment

There is no capital investment for setting up a computing environment. The investment can be used for core business needs.

Cost-Efficient

As the cloud services are charged based on a pay-per-use, the number of users can be planned and accordingly the payment is made. Hence, it proves to be cost efficient.

Infinite Computing Facility

As the computing facility is scalable and can be expanded on short notice as needed, an enterprise is powered with an infinite computing facility.

Easy Data Maintenance

Data maintenance becomes easy and similarly the back-up restore facility also becomes easy. Besides, cloud-service providers generally offer expert services as part of the cloud service on data maintenance.

Seamless Software Integration

In cloud services, software integration is seamless. Additionally, a feature of cloud is to select and choose only those software services that the enterprises need, which becomes again cost-effective for an enterprise.

Wider Access to Information

Accessing information through cloud becomes much easier as one needs to access data through the Internet unlike the private data centre, where one needs to be present on the premises or should deploy VPN for a secure access of data.

Speedier Implementation of Application Projects

As the cloud service is a ready-to-deploy environment, enterprises can speedily deploy application projects for the benefit of the enterprise operations.

Shortcomings

Cloud services are not without shortcomings.

These are detailed below:

Complex Technology Issues

Unlike a private data centre, where an enterprise sets up the hardware and enterprise architecture based on the most suitable options, cloud services are set up based on the technical optimisation as preferred by the cloud-service provider. Also, when accessed through the Internet, there are several possibilities of having technical challenges.

Security

All said and done, a cloud environment is accessed through a public Internet cloud, which is considered to be high-risk, even though an adequate technical-risk mitigation software can be set up.

Intrusions and Attack

As a cloud-service provider facility is accessed through open Internet, the environment is exposed to public Internet (though with adequate securities such as firewall, etc.); the cloud environment is thus prone to intrusion and hacking.

Mobile Computing

Mobile computing can be defined as the usage of portable devices such as laptops or any other handheld devices (PDAs, smartphones) in conjunction with the mobile communication technology to access enterprise systems from anywhere in the world.

Mobile Communication Infrastructure

Mobile infrastructure is about the system infrastructure that ensures a seamless continuous service of mobile communications. This includes switches, exchanges, protocol services and bandwidth. It also includes towers which receive and transmit signals in the air to and from mobile devices. These infrastructures are typically set up and maintained by the mobile communication service provider.

Mobile Hardware

Mobile hardware is comparable to the client-end hardware in client-server technology. Through mobile hardware, the user either communicates or performs computing activities. A mobile hardware could be any handheld device such as a mobile phone, smartphone, PDA, laptop, etc. A SIM card is an integral part of the mobile-phone hardware. Most of the mobile devices are fully duplex, which means that the device is capable of receiving and sending signals simultaneously.

Mobile Software

In order to run a computer, we will need an operating system as for laptops, PDAs, etc. In case of mobile and smartphones, one needs a mobile operating system, which is mobile software. Examples of mobile/smart phone operating systems are Apple iOS, Android, Windows mobile, Blackberry OS.

Mobile Phone Business Applications: Benefits and Concerns

One of the recent technology breakthroughs is the convergence of IT, communication technologies and entertainment technologies in a single device. As a result, mobile smart devices such as smartphones or tablet PCs are capable of accessing the enterprise's computing environment and one can perform transaction activities through the mobile device.

Mobile Computing: Benefits

Improve the Productivity of the Workforce: The mobile device's access to enterprise computing infrastructure offers the facilitation of employees to perform activities anytime anywhere. This dramatically improves the productivity of an employee as temporal and geographical barriers are removed through the mobile access facility. It also offers a better work–life balance to the employees.

Improved Customer Service: Mobile services to customers help in improving the customer service. For example, status alerts on customer orders on mobile, access to self-service CRM system through the mobile applications help customers.

IT Incident Support can be Speedier: As IT environment and application service have to be up and running 24/7, there is a need for an immediate fixing of issues/incidents, if any, as reported. Mobile access facilitates the IT-subject matter expert to log on from anywhere anytime that helps in fixing issues the speedily and easily.

Enhances Business Process Reengineering Benefits: In the past, the business process reengineering was carried out in terms of IT. Now with the mobile communication technologies' convergence with IT, BPR results have been enhanced. Examples of mobile application enhancing BPR benefits are a commercial bank offering mobile-banking facility to its customer and a cab-service provider offering cab booking through mobile device thus enhancing business results.

Processes Getting Executed beyond Geographical Boundaries: With mobile computing, the enterprise processes are getting executed beyond geographical boundaries and time, even if employees of the enterprises are travelling.

Mobile Computing: Concerns

Security: Web-based access through the Internet to the enterprise's IT resources poses major concerns as it exposes to risk. Because of mobile device access and the growth of mobile devices, mobile access to IT resources of an enterprise results in risks at an exponential level.

Misrepresentation through Valid Credentials: Like Internet access, mobile access is also user-validated through the credentials provided by the user. If a misrepresentation occurs—when a stranger logs on to the system through the valid credentials of a genuine customer—the stranger will be allowed by the system to access all IT resources. This is again a major security risk.

Dependency on Stored Power: Mobile devices work on stored power in the battery and have limitations on long duration usage.

Health Issues: Radiowave signals of mobile devices are capable of causing adverse impacts on the health of human beings and of society.

Network Virtualisation

In a local area network, there are several computing resources that are generally present. Capacities of all these resources may not be uniformly used or work may not be evenly loaded on these computing resources, as the nodes and servers may be assigned to dedicated purposes and the workload and capacity utilisation of these may not be even or uniform.

Network Virtualisation helps to create an abstract layer of all the computing resources such as memory, CPU, network, hard disk I/O and storage and provided for workload in a data centre, virtualised and pooled. As we are aware, a data centre is a physical location where there are several hundreds or thousands of installed and interconnected servers. This is accomplished through the network virtualisation software.

Application of Network Virtualisation

Server Consolidation

Through server consolidation, physical servers are pooled into the virtual network as guest resources and pooled server resources are provided for workload. This process ensures an optimal usage of all servers.

Disaster Recovery System

Every enterprise creates a disaster-recovery system, by having a replica of their existing environment with data to ensure that the disaster recovery becomes a back-up for any untoward incident that damages the primary/original computer environment. This is more towards risk mitigation as certain industries like banks totally depend on the computerised environment for running their business. Network virtualisation is applied in setting up a disaster-recovery environment to mirror the production environment.

Testing and Training

In order to judiciously use the investible resources, the enterprises will invest in a robust-production IT environment, where all the online real-time commercial transactions will be recorded. A production environment is set up after installing, configuring and developing the solutions in the development environment and replicating the development environment in another environment, where testing will be carried out and this is called testing environment; once development and testing are carried out, the testing environment will be mirrored in another environment called training environment, where end-users of a computerised solution are trained. End-users are, in some enterprises, allowed to freely log on to the training environment and perform notional transactions to familiarise themselves. Network virtualisation helps in pooling the resources in testing and training environments and to optimise the investible resources.

Portable Application Files

Certain portable applications need temporary files to be stored and accessed from a portable device, such as a removable drive. Installation of such files in the removable device and storing the same files on the virtual network drives help in accessing the portable applications from anywhere with a redirection of the removable drive access to a network drive access.

Portable Workspaces

This can be best explained with the help of Apple devices. For example, if a user is performing a computing activity using a Mac desktop at home, the same work can be continued at the stage where it has been left off after reaching office by logging on to Mac Book as the work can be saved on iCloud. iCloud in this case is acting as a virtual network providing virtual workspace.

Grid Computing

Grid computing is a concept in which all resources share the similar resources in a network. For example, in a local area network, if thousands of systems are connected by the same operating system, these individual systems share one's own resource with the others and similarly use others' resources. Grid computing is comparable to distributed computing. In a distributed-computing environment, all the users will share even the heterogeneous system resources, but in grid computing, in general, similar systems are shared as a collective resource by all users.

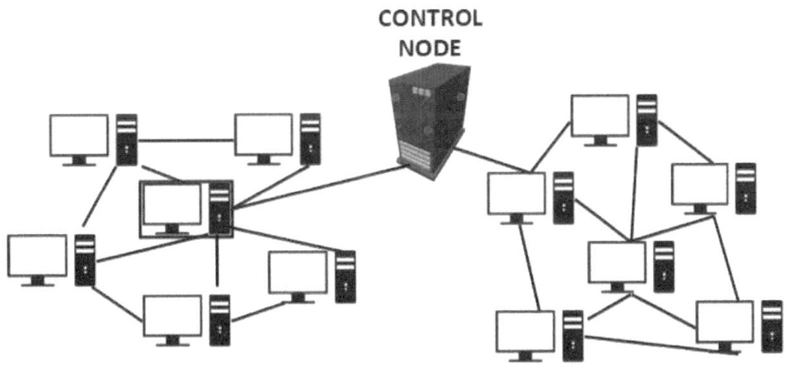

Diagram 5.8 Grid Computing

Usage of Grid Computing

Grid computing is very useful, where there is a huge data load to be handled. Some examples are given below:

Engineering Applications

In engineering applications, there will be a need for a huge data handling for simulation. In such scenarios, grid computing is very helpful.

Data-Mining Activities

In certain enterprises such as banks, retail, insurance, financial markets, etc. huge data is mined for performing analytical work; grid computing helps to utilise the resources on the grid.

High Volume e-Business Environment

e-Business environment will need a high volume of data handling and grid computing helps in this.

Big Data

Big Data is the latest concept in analytics. Hitherto, businesses were using commercial data generated within the enterprise to set up the intelligence systems. Now, in the interconnected world, business enterprises are attempting to collect data outside the enterprise such as from social networking sites and the general data available and non-commercial data such as images, videos, etc. for the intelligence set up. Besides the sources, enterprises have realised that the structured data is offering 20% of the information for decision-making and the remaining 80% is in an unstructured format such as call-centre transcripts, voice records, information in blogs, emails, social networking sites, data generated by sensors, etc. As the businesses wanted to tap the unstructured data, it was experienced that this data is being generated at a high ***'velocity'***, an exponentially high ***'volume'*** and is of a ***'variety*** of formats. Technologies have been developed for handling and processing this high-velocity, voluminous variety of this is called Big Data. Big Data benefits are reaped by high-volume transaction industries such as retail, telecom, financial services (including insurance), online e-Commerce industry, etc.

Under Big Data, conventional analytics approaches are completely changed. In conventional analytics, the analysts have a hypothesis—say, "Why profits are reduced? Perhaps increase in cost or reduction in revenue or both"—and look at the data for validating the hypothesis. Under the Big Data Analytics, the Data scientists will explore the data for gaining insights. The data scientists will explore the various data and pick up variables and gain insights by applying statistical methods or operational research concepts.

There are immense benefits gained by various businesses through the Big Data Analytics.

Two other important aspects of Big Data are

1. Predictive Analytics to predict the future events based on the analysis of available data.
2. Real-time analytics: Financial services and health-care industries extensively leverage Big Data Analytics for real-time analytics and spot decision-making.

Business Reporting Standards

As more and more government agencies embrace a computerised environment, they seek inputs from the public in a digital format. This information gathering is being attempted to be standardised across the world.

For example, for electronic data interchange—popularly known as **EDI**—there was **EDIFACT** (Electronic Data Interchange For Administration, Commerce and Transport).

Similarly, **XML** (eXtensible Markup Language) standards are used for commercial document standards.

Now, electronic governance compels governments to adopt certain standards. For example, XBLR is the standard set up by the Ministry of Company Affairs for companies to report their financial status as part of the regulatory reporting. **XBRL** (eXtensible Business Reporting Language) is a freely available and global standard for exchanging business information. XBRL allows the expression of semantic meaning commonly required in business reporting.

Benefits of Business Reporting Standards

1. Paperless lodgement of information.
2. Electronic record keeping, which is easy, convenient and less expensive.
3. Automatic filing of information, as on most of the occasions the system picks up data already existing in the system.
4. Easy to share the information.
5. Security authentication: In certain countries, a secure key authentication process ensures security to the electronic data/document. For example, AUSKey in Australia.
6. Instant validation: When the data is submitted electronically, the response for correct submission is received immediately.

Artificial Intelligence

The theory and development of computer systems able to perform tasks normally requiring human intelligence such as visual perception, speech recognition, decision-making and translation between languages are said to be Artificial Intelligence (or simply AI).

It is very complex subject, but being attempted by the computer industry to create a computer that can behave and respond like human intelligence. For example, IBM Watson is a kind of Artificial Intelligence; the Watson Lab of IBM had developed and experimented IBM Watson by making IBM Watson to compete with the other best participants in a quiz programme called *Jeopardy*. In this live programme, IBM Watson won against the other hitherto best participants.

If one analyses the prerequisites for an Artificial Intelligence to participate in a quiz programme, it calls for (1) comprehending the question (2) comprehending the context (3) searching for the answer (4) quick retrieval (5) competitive spirit (6) quick responsive action such as pressing the buzzer when ready with the answer and above all (7) speed in comprehension, contextual understanding, search for results, picking the right answer, action and communicating the answer clearly. These were considered achieved by the IBM Watson.

The fundamental difference between the human brain searches and response and computer searches and response is that the human brain is cognisant of context and relevance. By virtue of this attribute, Artificial Intelligence is sometimes called as Cognitive Computing.

Artificial Intelligence is attempting to achieve this and it is totally different from the other computer programming and application development, which repetitively performs the routine that has been coded. As against a standard computer's binary logic, Artificial Intelligence attempts to process the fuzzy logic that is comparable to human intelligence.

Information Handling—Applications, Responsiveness—Disciplines of and the Subject Covered by Artificial Intelligence

Artificial Intelligence calls for an appropriate responsiveness similar to that of the human mind which includes the broad disciplines of philosophy, psychology, numerical computing and computer science. With responsiveness in the disciplines, an Artificial Intelligence is expected to cover a vast volume of subjects such as reasoning, learning, planning, perception, knowledge acquisition, intelligent search, uncertainty management and predictability management. Combining responsiveness and subject knowledge, an Artificial Intelligence is expected to leverage a vast application area for offering answers or providing solutions to the question/challenges.

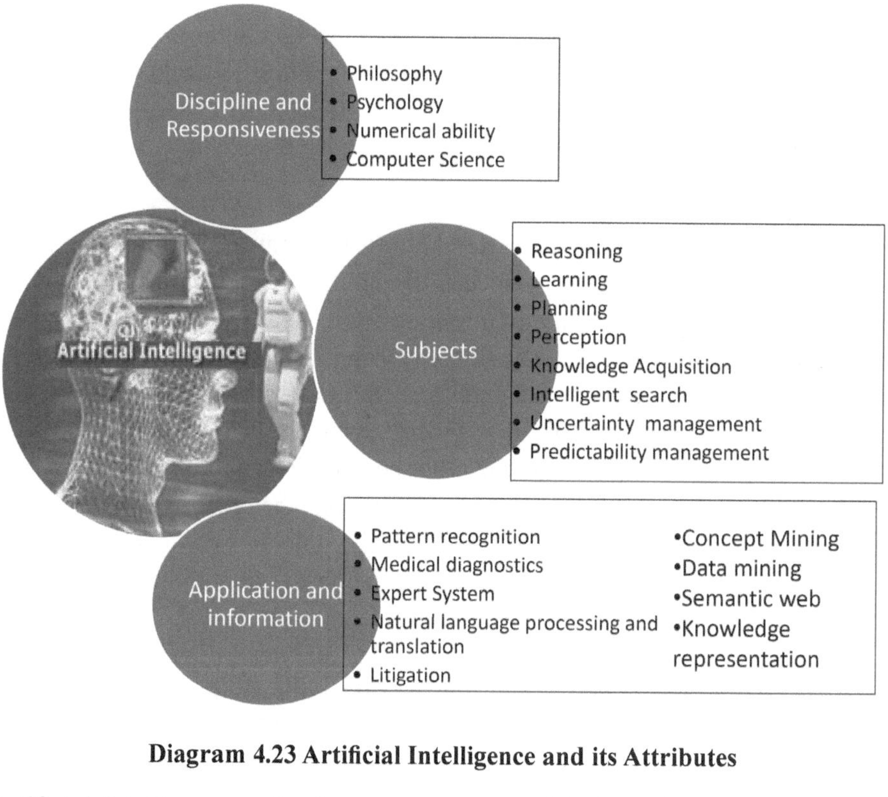

Diagram 4.23 Artificial Intelligence and its Attributes

An Artificial Intelligence calls for a vast amount of information handling, which can be termed as application. The application area can include: pattern recognition, medical diagnostics, expert systems, natural language processing and translation, litigation, concept mining, data mining, semantic web and knowledge representation.

Emergence of 'Smart World'

The combined effect of Internet technologies, computer science and Artificial Intelligence enabled the invention and innovation of various new devices such as smartphone, Smart Watch and Google Glass. These devices make the world smarter.

For example, a smartphone locates your geographical location based on the in-built GPS and maps. It has your official details such as your office phone #, official e-mail ID, etc. as you have already configured this information. It has your calendar details and agenda for the day.

Applications installed in your phone such as Google help you to become smarter by configuring—reminders, flight status, birthdays, topic updates, events nearby, news updates, next appointment, flights schedules, public alerts, public transit, sports, stocks, traffic, weather and so on.

When you swipe down your smartphone (iPhone) and tap the 'Today' tab, it tells you the day, date, weather forecast, calendar events, reminders for the day, stock quotes, the following day schedule, etc.

In addition, at the appropriate time, the phone provides alerts about the upcoming events to be attended, reminders of the activities to be carried out, etc. When a meeting is scheduled in a venue other than your current location, it tells how much time it is going to take for you to drive to the meeting venue, given the traffic and weather conditions.

Similar to making an individual smarter, every possible thing is made smarter through technologies. This is resulting in smarter home appliances, smarter automobiles, smarter electricity grids, smarter cities, smarter countries, smarter continents and eventually we are moving towards a smarter world.

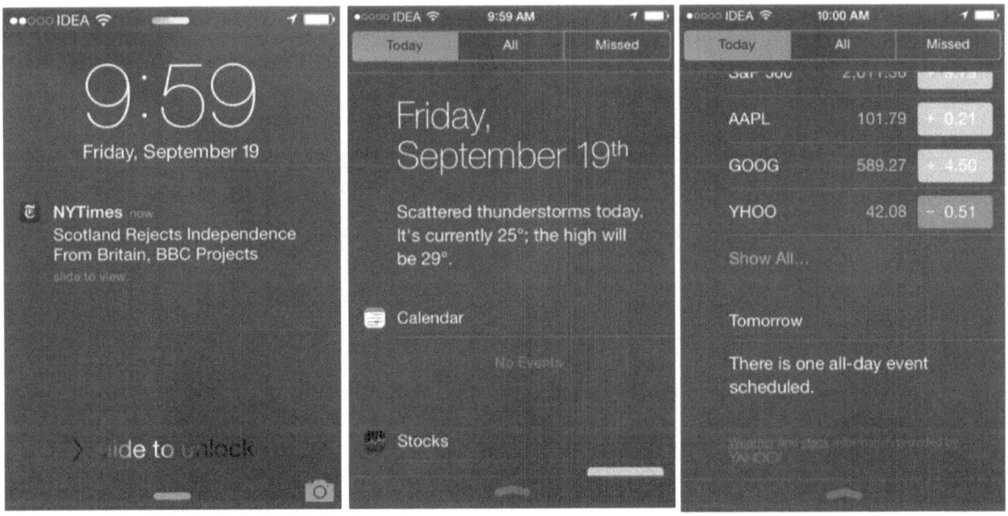

Diagram 4.24 Smartphone Assistance on 'Today' Status

Internet of Things

Internet is where computers are interconnected through the World Wide Web (WWW). Imagine that in addition to computers, various other electronic things get interconnected through the Internet. This is known as the Internet of Things (also known as IoT).

Some of the existing IoT applications currently in vogue are

1. Home appliances connected with each other and to a smartphone
2. Technical equipments in the field connected through the Internet to a system in the enterprise located in a common facility. For example, electric gridline equipments are connected together with a centralised system of the utility company for gathering data for optimisation and in response to the weather conditions.

There are several similar applications which are existing.

One of the researches that is currently (at the time of the writing of this book) taking place is the development of Intelligent Cars, where the cars are driven automatically, without a human driver and the cars are connected with one another through sensors to avoid accidents.

It is believed that by 2050, 50 billion things will be interconnected as part of IoT.

IoT is expected to have a wider application in the creation and management of Smarter Cities.

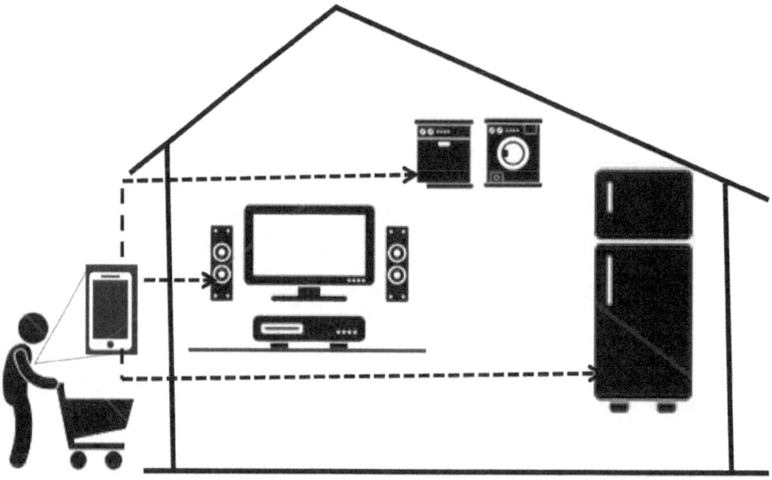

Diagram 4,25 Smart-Home Appliances and Smart Home

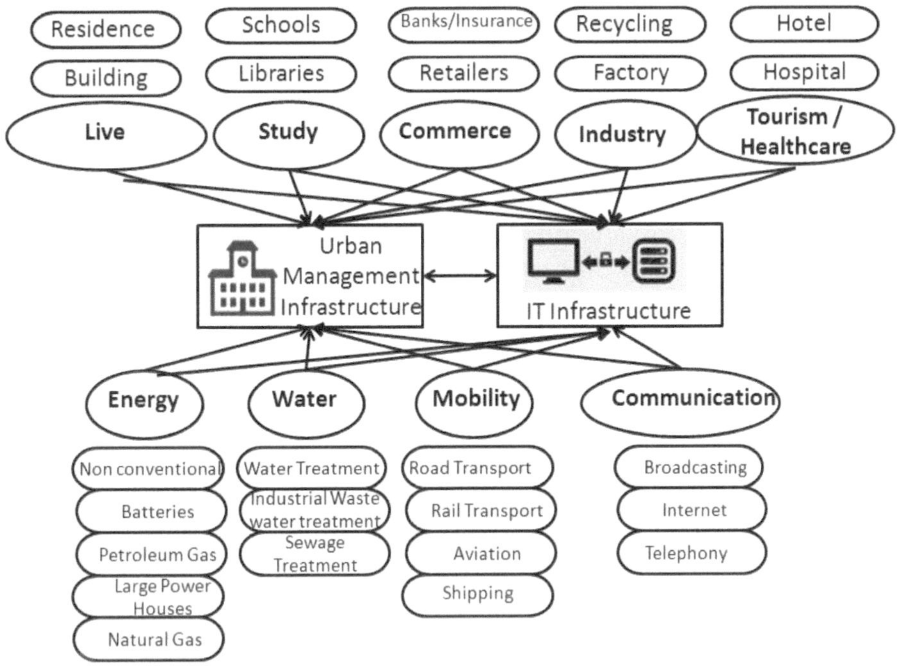

Diagram 4.26 Smarter Cities and Smarter World

One cannot underestimate the impact of IoT on the business and consequently in accounting and the various valuations in an enterprise. The primary reason is that through IoT, different data that is going to flow into an enterprise system is likely to influence the money value posted in the financial books of the records of the enterprise.

There will be a strong need for studying the data and the information flowing into the financial system in order to form an opinion or judgement about the correct and fair values stated in the enterprise books of accounts.

Wearable

Wearable computers are called as 'Wearable'. Wearable are comparable to some of the accessories such as a pair of spectacles or a watch we wear in our normal life. But in case of Wearable the difference is that the Wearable has certain electronic capabilities based on the features built-in and in addition these devices are capable of connecting to the Internet.

Wireless Personal Area Network (WPAN) and Wireless Body Area Network (WBAN) are the two interfacing Network standardisations.

Two most popular Wearable devices are

1. Google Glass

Google Glass is a head-mounted Wearable comparable to a pair of spectacles. Google Glass has the features of touchpad, camera and display built on the temple-frames.

Google applications such as Google Maps, Gmail and Google+ are available in Google Glass in its initial version. Subsequent versions are to include Facebook, Twitter and facial recognition software.

2. Apple Watch

It is a smartwatch invented by Apple Inc. Besides the functionality of a wrist watch, Apple Watch has fitness and health-oriented data collection features and has an integration capability with iOS. Through sensors such as Accelerometer, Gyroscope, Barometer, the Apple watch collects data on the human physiological system of the individual who wears the Apple Watch. It has the capability to connect to WiFi and Bluetooth. Apple Watch is expected to revolutionise the health-care monitoring of an individual who wears it through the health-care data collection and in providing it to the individual who wears the Apple Watch. Apple is also planning to integrate the 'Apple Pay' feature of payment through credit cards in the Apple Watch.

While Wearable is currently limited to personal application access and personal health-care monitoring, in future Wearable has the potential to be applied in accessing enterprise applications. Therefore, access, identity management and security become critical attributes for a safe Wearable.